Springer Theses

Recognizing Outstanding Ph.D. Research

Aims and Scope

The series "Springer Theses" brings together a selection of the very best Ph.D. theses from around the world and across the physical sciences. Nominated and endorsed by two recognized specialists, each published volume has been selected for its scientific excellence and the high impact of its contents for the pertinent field of research. For greater accessibility to non-specialists, the published versions include an extended introduction, as well as a foreword by the student's supervisor explaining the special relevance of the work for the field. As a whole, the series will provide a valuable resource both for newcomers to the research fields described, and for other scientists seeking detailed background information on special questions. Finally, it provides an accredited documentation of the valuable contributions made by today's younger generation of scientists.

Theses are accepted into the series by invited nomination only and must fulfill all of the following criteria

- They must be written in good English.
- The topic should fall within the confines of Chemistry, Physics, Earth Sciences, Engineering and related interdisciplinary fields such as Materials, Nanoscience, Chemical Engineering, Complex Systems and Biophysics.
- The work reported in the thesis must represent a significant scientific advance.
- If the thesis includes previously published material, permission to reproduce this must be gained from the respective copyright holder.
- They must have been examined and passed during the 12 months prior to nomination.
- Each thesis should include a foreword by the supervisor outlining the significance of its content.
- The theses should have a clearly defined structure including an introduction accessible to scientists not expert in that particular field.

More information about this series at http://www.springer.com/series/8790

Christopher H. T. Lee

Design, Analysis and Application of Magnetless Doubly Salient Machines

Doctoral Thesis accepted by
The University of Hong Kong, Hong Kong, China

 Springer

Author
Dr. Christopher H. T. Lee
Research Laboratory of Electronics
Massachusetts Institute of Technology
Cambridge, MA
USA

Supervisors
Prof. K. T. Chau
Department of Electrical
  and Electronic Engineering
The University of Hong Kong
Hong Kong
China

Prof. C. C. Chan
Department of Electrical
  and Electronic Engineering
The University of Hong Kong
Hong Kong
China

ISSN 2190-5053   ISSN 2190-5061   (electronic)
Springer Theses
ISBN 978-981-13-4991-1   ISBN 978-981-10-7077-8   (eBook)
https://doi.org/10.1007/978-981-10-7077-8

Printed on acid-free paper

This Springer imprint is published by Springer Nature
The registered company is Springer Nature Singapore Pte Ltd.
The registered company address is: 152 Beach Road, #21-01/04 Gateway East, Singapore 189721, Singapore

To Gogo

Supervisors' Foreword

The Confucian classic Li Ji, literally the Book of Rites, is a collection of articles written by Confucius's students and Confucian scholars during the Warring States Period (475–221 BC) in ancient China, which not only describes rules and regulations, but also benevolence and morality. Among 130 articles in Li Ji, the article Xue Ji is influential in education. Its essence—Teaching others can gain half of the effect of learning—is the teaching philosophy that I consider as the doctrine. Being a Ph.D. supervisor of Christopher, I can fully testify the essence of Xue Ji— Whenever I teach him, I can find my limitation in knowledge.

This thesis describes the scientific achievements of Dr. Christopher Ho Tin Lee, who has completed his doctoral program in Department of Electrical and Electronic Engineering, The University of Hong Kong. Dr. Lee has achieved excellent research outcome on the electric machine design, particularly on magnetless machine topologies, during his Ph.D. studies. As his Ph.D. supervisor, I would like to introduce two important findings from his doctoral work.

One of the most important findings from his works is the development of the electronic-geared (EG) machine. By purposely combining the design philosophies of two extreme machine types, the resulting EG machine can switch between two modes of operation to cater for a wide range of situations. Consequently, the undesirable mechanical gearbox can be eliminated. This particular feature is highly desirable for industrial applications, such as wind power generation and electric vehicle (EV) propulsion. Another important finding comes from the innovation of magnetic steering (MS) concept. Upon the installation of MS windings in the double-rotor machine, the resulting double-rotor machine can enable EVs to perform stable curvilinear movement. Because of the magnetic interlocking mechanism provided by the MS windings, this machine can provide higher stability and better reliability than the existing systems.

The findings in this thesis have been published in top engineering journals, namely the *IEEE Transactions on Industrial Electronics, IEEE Transactions on*

Energy Conversion, and *IEEE Transactions on Magnetics*. I hope these research works would stimulate more researchers to work on electric machines.

Hong Kong, China   Prof. K. T. Chau
October 2017

Sustainable energy and environment protection are among current major global grand challenges. As one of the most promising solutions to these issues, the developments of the wind power generation and electric vehicle (EV) have been speeding up recently. As the key part of these emerging applications, the design and analysis of the electric machines have become a hot research topic.

Because of the high power and torque performances, the permanent-magnet (PM) machines have dominated the industrial market in the past few decades. However, due to the limited and fluctuating supply of the PM materials, the construction costs of the PM machines have been raised tremendously. Consequently, without installation of any PM materials, the magnetless machines have become very attractive recently. The thesis of Dr. Christopher Ho Tin Lee provides a comprehensive study of the magnetless machines, particularly for the wind power generation and EV applications. Throughout the doctoral studies, Dr. Lee has proposed some interesting ideas, such as electronic-geared operation and magnetic interlocking mechanism, to improve the overall performances of the magnetless machines.

Dr. Lee is a hardworking and self-motivated student, and I enjoy supervising him towards his doctorate. He has shown his talents and dedication towards his research. I hope you will enjoy reading his work, as much as I do.

Hong Kong, China   Prof. C. C. Chan
October 2017

Abstract

Owing to growing concerns on energy utilization and environmental protection, research on electric vehicle (EV) and wind power generation has drawn much attention in the past few decades. As a key component of these applications, the development of electric machines has become a hot research topic since the last century. While the permanent-magnet (PM) brushless machines have dominated the industrial and domestic markets for many years due to their outstanding performances when compared to their counterparts, the manufacturing costs of the PM machines have tremendously increased lately as a result of virtual market monopoly and fluctuating supply of PM materials. Therefore, with the absence of PM materials, the advanced magnetless doubly salient brushless machines that provide great cost-effectiveness have become more popular recently.

Without the need to install high-energy-density PM materials, magnetless doubly salient machines undoubtedly suffer from relatively lower torque densities when compared to their PM counterparts. To resolve this problem, the development of torque improving techniques has become an interesting research topic for the magnetless machines. On the other hand, unlike the PM machines whose magnetic fluxes are uncontrollable, the advanced magnetless machines can utilize its independent DC-field excitation for flux regulation. Hence, the operating ranges of the magnetless machines can be effectively extended.

The purpose of this thesis is to investigate the characteristics of existing magnetless doubly salient machines, analyze their design philosophies and propose new topologies for various applications. Firstly, the background of magnetless machines and previous works conducted by other researchers are introduced. Based on the study of the existing works, the upcoming trend and the potential development are also reviewed. Secondly, the torque improving structures, namely the multi-tooth structure, the double-rotor (DR) structure, the singly fed mechanical-offset (SF-MO) structure, the flux-reversal (FR) and the axial-field (AF) structure, that are purposely implemented into the magnetless machines are discussed. Next, the magnetic steering (MS) machine, which utilizes the concept of magnetic interlocking to achieve higher stability for curvilinear motion, is proposed. Then, the idea to incorporate the design philosophies of the two different machines to form new dual-mode machines is covered. With the reconfiguration of winding arrangement and the support from controllable DC-field excitation, the proposed machines can further extend their operating range to fulfill extreme conditions required by the applications of EV and wind energy harvesting. Finally, all the key performances of the proposed machines are thoroughly analyzed by the finite element method (FEM), while the experimental setups have also been developed to verify the proposed concepts.

Parts of this thesis have been published in the following journal papers:

[J1] **C. H. T. Lee**, K. T. Chau, C. Liu, and C. C. Chan, "Development of a singly fed mechanical-offset machine for electric vehicles," *IEEE Transactions on Energy Conversion*, accepted.

[J2] **C. H. T. Lee**, K. T. Chau, C. Liu, and C. C. Chan, "Overview of magnetless brushless machines," *IET Electric Power Applications*, accepted.

[J3] **C. H. T. Lee**, K. T. Chau, and L. B. Cao, "Development of reliable gearless motors for electric vehicles," *IEEE Transactions on Magnetics*, vol. 53, p. 6500308, November 2017. (Invited Paper)

[J4] **C. H. T. Lee**, K. T. Chau, and C. Liu, "Design and analysis of an electronic-geared magnetless machine for electric vehicles," *IEEE Transactions on Industrial Electronics*, vol. 63, no. 11, pp. 6705–6714, November 2016.

[J5] **C. H. T. Lee**, K. T. Chau, and C. Liu, "Design and analysis of a cost-effective magnetless multi-phase flux-reversal DC-field machine for wind power generation," *IEEE Transactions on Energy Conversion*, vol. 30, no. 4, pp. 1565–1573, December 2015.

[J6] **C. H. T. Lee**, K. T. Chau, and C. Liu, "Design and analysis of a dual-mode flux-switching doubly salient DC-field magnetless machine for wind power harvesting," *IET Renewable Power Generation*, vol. 9, no. 8, pp. 908–915, April 2015.

[J7] **C. H. T. Lee**, K. T. Chau, C. Liu, T. W. Ching, and F. Li, "Mechanical offset for torque ripple reduction for magnetless double-stator doubly-salient machine," *IEEE Transactions on Magnetics*, vol. 50, no. 11, p. 8103304, November 2014.

[J8] **C. H. T. Lee**, C. Liu, and K. T. Chau, "A magnetless axial-flux machine for range-extended electric vehicle," *Energies*, vol. 7, no. 3, pp. 1483–1499, March 2014.

[J9] **C. H. T. Lee**, K. T. Chau, and C. Liu, "Electromagnetic design and analysis of magnetless double-rotor dual-mode machines," *Progress In Electromagnetics Research*, vol. 142, pp. 333–351, November 2013.

[J10] **C. H. T. Lee**, K. T. Chau, C. Liu, D. Wu, and S. Gao, "Quantitative comparison and analysis of magnetless machines with reluctance topologies," *IEEE Transactions on Magnetics*, vol. 49, no. 7, pp. 3969–3972, July 2013.

Acknowledgements

Giving thanks always and for everything to God the Father in the name of our Lord Jesus Christ

Ephesians 5:20

I would first like to thank my thesis supervisors, Prof. K. T. Chau and Prof. C. C. Chan for their precious guidance, assistance and encouragement over the past years. The office door of Prof. Chau is always open whenever I come across any brain-stormed ideas or any trouble spots about my research. He consistently provides me freedom to swim in the sea of the knowledge, while steers me to the right direction whenever necessary. In the meantime, Prof. Chan has shown me the true meaning of life, where his full dedication, passion and commitment have inspired me a lot. Should I not be the student and follower of both excellent teachers, I shall not develop myself academically and spiritually.

I must deliver my gratitude to Mr. Raymond Ho, the technician of our research team. Without his tireless and professional assistance, the experimental works of my research project could have never been accomplished.

I would also like to express my indebtedness to my colleagues and all whom I have worked with, most notably Dr. C. Liu, Dr. W. Li, Dr. D. Wu, Dr. S. Gao, Dr. Z. Zhang, Dr. F. Li, Dr. M. Chen, Dr. T. W. Ching, Dr. F. Lin, Miss H. Fan, Mr. C. Jiang, Mr. W. Han and Mr. L. Cao for their patients and genuine supports.

Last but not least, expressions of gratitude and apology are directed to my mother, Linda, my brother, Gary and my fiancée, Gogo, who patiently endure the long working hours of my work. Without their complete understanding and support, this work would have been much more difficult.

Contents

List of Figures

List of Tables

Chapter 1
Introduction

1.1 Background of Electric Machines

As one of the most efficient energy conversion devices that interconvert the mechanical and electrical powers, the electric machines have been employed in various industrial and domestic applications for more than one hundred years. To be specific, the electric machines have consumed about 65% of total generated electrical energy in the developed countries [1], and this particular invention has been renowned as one of the most important milestones for modern civilization. Generally speaking, there are two major electric machines available, namely the brushed machine and the brushless machine. By the classification based on the employed materials, the brushless machine can be further divided into two subgroups, namely the permanent-magnet (PM) brushless machine and the magnetless brushless machine. Most of the major machine types, including the brushed, the brushless, the PM brushless and the magnetless brushless machines have been classified as shown in Fig. 1.1.

Because of the simple control scheme and wide operating range, the brushed machines have drawn most of the attentions in the early years so that various topologies have been proposed [2]. However, the brushed machines suffer from the unavoidable frictional losses due to the commutator operation and hence the overall stability is degraded. To maintain the normal functionality of the brushed machines, the undesirable regular maintenances are needed. On the other hand, without installation of any brushes and commutators, the brushless machines can eliminate the profound maintenance problem that exists in the brushed machines. However, these types of machine were not popular because of the complex control algorithms and expensive material costs of power electronics [3].

The brushed machines have dominated the industrial and domestic markets for a long period of time, yet the situation has changed since the advancement of the power electronic technologies. With the development of the power electronics, the complex control algorithms for the brushless machines can be achieved at a very

© Springer Nature Singapore Pte Ltd. 2018
C. H. T. Lee, *Design, Analysis and Application of Magnetless Doubly Salient Machines*, Springer Theses,
https://doi.org/10.1007/978-981-10-7077-8_1

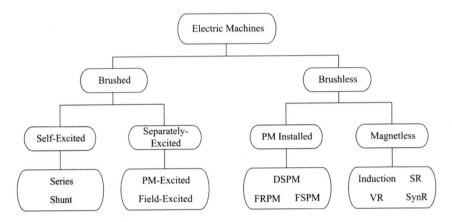

Fig. 1.1 Classification of the modern electric machines

low cost [4]. Therefore, there is an increasing trend to employ the brushless machines for various applications, while the brushed machines are fading out. Nowadays, the brushless machines have served as the predominated candidates for many industrial applications, such as the electric vehicle (EV) application and the wind power generation.

1.2   Development of Brushless Electric Machines

Owing to the ever enhancing concerns on environmental protection as well as energy efficiency, the development of EV application and wind power generation has been speeding up [5, 6]. Served as the key element for these applications, the electric machines, particularly the brushless one, have drawn many attentions in the past few decades. Indeed, the concept of the brushless machines has been probably first proposed by Nikola Tesla in the late 1800s [7]. It should be noted that the brushless machine was originally developed based on the magnetless structure, i.e., without any PM components. Due to the robust and simple structure, the magentless brushless machines have been more attractive than the PM one. Yet, the leading position of the magnetless brushless machines has been challenged by the high-performance PM brushless machines that were proposed in 1950s [8].

By employing the high-energy-density PM materials, the PM brushless machines can offer better performances, including higher efficiency and higher power density, as compared with the magnetless counterparts. Hence, the PM brushless machines have started to overtake the leading role for the industrial applications. Upon the enhancing demands on the environmental protection, the developing rate of the PM brushless machines has been accelerating so that numerous topologies have been proposed in the last few decades. The PM brushless machines can be categorized based on its PM allocation and the flux-linkage characteristic. To be specific, there are generally three major types available, namely the doubly salient permanent-magnet (DSPM) type [9], the flux-reversal permanent-magnet (FRPM) type [10] and the flux-switching permanent-magnet (FSPM) type [11].

Even though the PM brushless machines can offer the outstanding performances, these types of machines have suffered from the high material costs in the past few years [12]. Apart from the high construction costs, the PM flux density cannot be controlled effectively so that the operating ranges of the PM machines are limited. All these shortcomings have hindered the further development of the PM machines, and hence the researchers and engineers have instead started to pay more attentions on the magnetless candidates recently. Since the very first magnetless doubly salient brushless machine, the induction machine was proposed in more than one century ago, some other magnetless doubly salient brushless candidates have been proposed throughout the years. In particular, there are four fundamental magnetless brushless machines available, namely the induction machine, the synchronous reluctance (SynR) machine, the Vernier reluctance (VR) machine and the switched reluctance (SR) machine.

1.3 Research Objectives

The main research objective of this project is to develop and analyze the new types of magnetless doubly salient brushless machines for industrial applications. In particular, the project aims to achieve the goals as follows

- To design and analyze the new magnetless doubly salient brushless machines with improved torque performances and higher cost-effectiveness.
- To quantitatively compare the key performances between the proposed machines and the commonly employed candidates.
- To explore and suggest the new design philosophies for the magnetless machines in order to extend the operating range and control flexibility.
- To develop and implement the experimental setups of the proposed magnetless machines for concept verifications.
- To apply the proposed magnetless machines into various potential applications, such as EV application or wind power generation.

1.4 Outline

This thesis consists of eleven chapters, where several sections and subsections are included. To improve the readability, the chapters are purposely allocated into three distinctive parts, namely (i) Part I: Modeling, numerical analysis and experimental verification of torque improving topologies; (ii) Part II: Design, analysis and application of advanced magnetless machines on wind power generation; and

(iii) Part III: A comprehensive study of advanced magnetless machines on electric and hybrid vehicles. An outline of all chapters is given as follows

In this Chapter, the background and the current status of the commonly employed electric machines are described and discussed, and hence the research objectives are defined.

In Chap. 2, the introduction, modern developments and upcoming technologies of the magnetless brushless machines are reviewed.

Part I: Modeling, numerical analysis and experimental verification of torque improving topologies

In Chap. 3, the torque improvement design structure, namely the multi-tooth structure has been analyzed and utilized to develop the new magnetless machines. In the meantime, the new torque ripple minimization topology, namely the mechanical-offset (MO) topology is proposed.

In Chap. 4, the concept of double-rotor (DR) structure is newly implemented into the magnetless machine topologies, leading to form the flexible machines with two sets of independently driven shafts, purposely for the special direct-drive applications.

In Chap. 5, based on the foundation in Chap. 3, the concept of electrical-offset (EO) is implemented into the MO machine, and hence forming the singly fed mechanical-offset (SF-MO) machine. The propose SF-MO machine not only can minimize its torque ripple to a very desirable level, but also reduce the costs of power electronics and control complexity.

Part II: Design, analysis and application of advanced magnetless machines on wind power generations

In Chap. 6, the stator pole modulation technique is newly proposed to form the cost-effective flux-reversal DC-field (FRDC) machine. Based on the proposed modulated stator design, the DC-field winding can then be wound in a way to perform similarly as the FRPM machine does. With the bipolar flux-linkage characteristic, the proposed FRDC machine can offer satisfactory torque performances with attractive cost-effectiveness.

In Chap. 7, the two machine design philosophies, namely the DSDC and the FSDC machines are incorporated together, to form the new dual-mode machine, purposely for the wind power harvesting applications. On the employment of different winding configurations, the proposed machine can switch between two different modes to behave similarly as the corresponding machines do. Therefore, the proposed machine can offer higher flexibility and stability to cater different possible wind situations.

Part III: A comprehensive study of advanced magnetless machines on electric and hybrid vehicles

In Chap. 8, the axial-field (AF) structure is newly proposed to implement with the doubly salient DC-field (DSDC) machine, to form the high-performance AF-DSDC machine. With its superior torque density performance and flux-weakening characteristics, the proposed machine is highly suitable for the ranged-extended electric vehicle (RE-EV) applications.

In Chap. 9, the magnetic steering (MS) machines is proposed to realize the magnetic differential (MagD) system for EV applications. The concept of magnetic interlocking is newly testified to achieve differential action with higher reliability for curvilinear motion.

In Chap. 10, based on the foundation in Chap. 7, a new dual-mode electronic-geared (EG) machine is developed, purposely for the high-performance EV applications. To be specific, the design criteria of the back electromotive force (EMF) waveform and the winding arrangement are thoroughly analyzed. Hence, the proposed EG machine can offer the smoother torque at the low-gear situation, and the better torque density at the high-gear situation.

Finally, in Chap. 11, the summary and conclusions of this thesis are given. The contributions of this project are highlighted, while the suggestions for possible improvement are also provided.

References

1. B.K. Bose, Power electronics and motion control-technology status and recent trends. IEEE Trans. Ind. Appl. **29**(5), 902–909 (1993)
2. S.J. Chapman, *Electric Machinery Fundamentals* (McGraw-Hall Press, New York, 2003)
3. P.C. Sen, Electric motor drives and control-past, present and future. IEEE Trans. Ind. Electron. **37**(6), 575–592 (1990)
4. T.A. Lipo, Recent progress in the development of the solid-state AC motor drives. IEEE Trans. Power Electron. **3**(2), 102–117 (1988)
5. K.T. Chau, *Electric Vehicle Machines and Drives – Design, Analysis and Application* (Wiley-IEEE Press, London, 2015)
6. S. Muller, M. Deicke, R.W. De Donchker, Doubly fed induction generator systems for wind turbines. IEEE Ind. Appl. Mag. **8**(3), 26–33 (2002)
7. N. Tesla, A new system of alternate current motors and transformers. Trans. Am. Inst. Electr. Eng. **5**(V), 308–324 (1888)
8. S.E. Rauch, L.J. Johnson, Design principles of flux-switch alternators. Trans. Am. Inst. Electr. Eng. **74**(3), 1261–1268 (1955)
9. Y. Liao, F. Liang, T.A. Lipo, A novel permanent magnet motor with doubly salient structure. IEEE Trans. Ind. Appl. **31**(56), 1069–1078 (1995)
10. R.P. Deodhar, S. Andersson, I. Boldea, T.J.E. Miller, The flux-reversal machine: a new brushless doubly-salient permanent-magnet machine. IEEE Trans. Ind. Electron. **33**(4), 925–934 (1997)
11. E. Hoang, A.H. Ben-Ahmed, J. Lucidarme, Switching flux permanent magnet polyphasic machines, in *Proceeding of Europe Conference Power Electronic Application*, Trondheim, Norway, pp. 903–908, September 1997
12. I. Boldea, L.N. Tutelea, L. Parsa, D. Dorrell, Automotive electric propulsion systems with reduced or no permanent magnets: An overview. IEEE Trans. Ind. Electron. **61**(10), 5696–5711 (2014)

Chapter 2
Overview of Magnetless Doubly Salient Brushless Machines

2.1 Introduction

Due to a large variety of industrial applications, including the electric vehicles (EVs), wind power generations, ship propulsion systems and robotic applications, the development of the electric machines has been a hot research topic in the last century [1]. The brushed machines and the brushless machines are the two major types of the electric machines [2]. As compared with the brushed one, the brushless machine can enjoy the absolute advantage of maintenance-free operation, so that this type of machines has become the major trend since the last few decades [3]. By the classification of the employed materials, the brushless machine can be further divided into two subgroups, namely the permanent-magnet (PM) machines and the magnetless machines.

Upon the installation of the high-energy-density PM materials, the PM candidates can provide superior performances and hence this type of machines has become very attractive for various applications. However, due to the virtual market monopoly and fluctuation of supply, there is drastically increase of PM material costs [4]. To increase the product competitiveness as well as the market penetration, the magnetless machine with higher cost-effectiveness has become more popular recently.

The purpose of this chapter is to provide an overview of the magnetless doubly salient machines. Therefore, the current technologies of the magnetless machines, including machine structures, characteristics, operation principles, control algorithms and upcoming trends will be reviewed and discussed.

© Springer Nature Singapore Pte Ltd. 2018
C. H. T. Lee, *Design, Analysis and Application of Magnetless Doubly Salient Machines*, Springer Theses,
https://doi.org/10.1007/978-981-10-7077-8_2

2.2 Background of Magnetless Brushless Machines

The magnetless brushless machines has been probably first developed in late 1800s [5], when the first brushless alternator machine was proposed by Nikola Tesla. Without the installation of any PM materials, the magnetless machine candidate enjoys the definite merits of cost-benefit as well as robustness structure, so that it has dominated the industrial market for a long period of time. Yet, the situation has changed in 1950s, when the high-performance PM brushless machines were proposed [6].

Upon the new design structures and construction techniques, the PM machines that provide the relatively high power densities have therefore taken the leading position over the magnetless machines in the past few decades. In particular, the doubly salient PM (DSPM) machine [7], the flux-reversal PM (FRPM) machine [8] and the flux-switching PM (FSPM) machine [9] are the major PM candidates employed in various applications. Nevertheless, the supply of the PM materials is fluctuating, and hence the construction costs of the PM machines have risen drastically. Therefore, the development of the magnetless machines has once again become very active recently.

2.3 Well-Developed Topologies, Features and Performances

In general, there are four major types of fundamental magnetless brushless machines available, namely the induction machine [5], the synchronous reluctance (SynR) machine [10], the Vernier reluctance (VR) machine [11] and the switched reluctance (SR) machine [12]. Even though all of these candidates are categorized as magnetless brushless machines, each of them exhibits the unique features and distinct characteristics.

2.3.1 Induction Machine

The induction machine, which was first proposed by Nikola Tesla in 1888, has been regarded as one of the most developed electric machines [13]. Literally speaking, the induction machine is operated based on the principle of electromagnetic induction, which can be achieved by the rotating magnetic field. The major type of the induction brushless machine consists of the squirrel cage structure [14], as shown in Fig. 2.1. The basic performances of the induction machine can be accurately estimated by the equation modeling [15].

Fig. 2.1 Induction machine with squirrel cage structure [14]

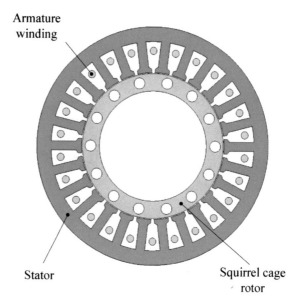

Armature winding

Stator

Squirrel cage rotor

Even though the three-phase topology is the most common design, the induction machines with higher number of phases have also drawn some attentions for various applications [16]. Apart from the design of machine structure, the control algorithm is another attractive technology that has drawn considerable attentions [17]. In particular, a huge variety of sensorless control schemes have been employed in the industrial applications [18].

2.3.2   Synchronous Reluctance Machine

As one of the earliest types of electric machines, the SynR machine has been developed as a cylindrical rotor with multiple slits in early 1900s [10]. The SynR machine is operated based on the principle of minimum reluctance, which can be achieved by the rotating magnetic field. The major torque component of the SynR machine comes from the reluctance torque, which is directly proportional to the saliency ratio. To improve the saliency ratio as well as the output torque performance, a segmental rotor structure [19] and an axially laminated rotor structure [20] have been developed. The optimization of the rotor design has drawn many attentions [21], while the most common topology is shown in Fig. 2.2.

Apart from the average torque value, the torque ripple issue is another important criterion to determine the machine performance. To minimize the torque pulsation, the asymmetrical rotor barrier arrangement [22] and the torque harmonic compensation approach [23] have been proposed.

Fig. 2.2 Synchronous
reluctance machine [21]

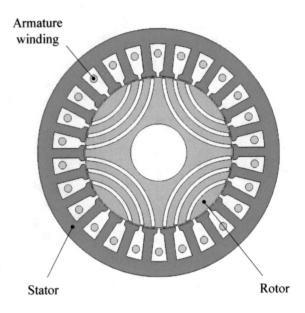

Armature
winding

Stator   Rotor

2.3.3 Vernier Reluctance Machine

Same as the SynR machine, the VR machine is also operated based on the principle
of minimum reluctance. Yet, the VR machine distinguishes itself by the salient
rotor pole structure [24], as shown in Fig. 2.3. In the meantime, the VR machine is

Fig. 2.3 Low-speed Vernier
reluctance machine [24]

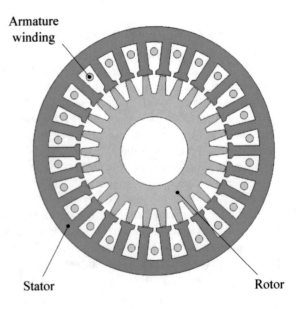

Armature
winding

Stator   Rotor

operated based on the rotating field scheme [25], where the operation is similar as the SynR machine.

Based on the rotating field operation with the salient pole structure, the relatively smoother torque can be produced even in the low-speed situation. To utilizing these characteristics, the VR machine is purposely designed with the machine structure of relatively higher number of poles, so that the high-torque low-speed operation can be achieved [26]. The VR machine has been extended to form the double-stator VR machine [27] and the double-rotor VR machine [28]. Yet, same as other the reluctance machines, the VR machine also suffers from the problem of relatively lower power factor. To improve the power factor, the VR machine with the doubly-slotted stator structure has been proposed [29].

2.3.4   Switched Reluctance Machine

Same as the VR machine, the SR machine consists of the salient pole structure and is also operated based on the principle of minimum reluctance. Yet, unlike the SynR and VR machines that are operated by the rotating field, the SR machine is instead operated by the phase current pulse scheme. The simplest SR machine consists of the one-phase 2/2-pole (two-stator-pole/two-rotor-pole) topology [30], while it is not practical for industrial applications. The SR machine can be further developed as the three-phase 6/4-pole structure [31], as shown in Fig. 2.4, the three-phase 12/8-pole structure [32], the three-phase 24/16-pole structure [33] and the four-phase 8/6-pole structure [30]. Recently, the new topology with the higher ratio of rotor to stator pole

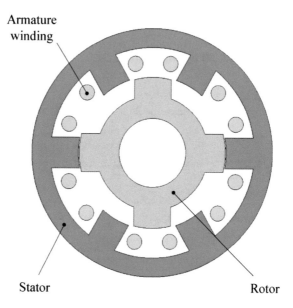

Fig. 2.4 Three-phase 6/4-pole switched reluctance machine [31]

Armature winding

Stator

Rotor

has also been explored [34]. With the new stator to rotor pole arrangement, more winding slot area as well as larger torque density can be achieved.

To improve the performance of the SR machine, the reduction of torque ripple and acoustic noise have become a hot research topic. The torque pulsations as well as the acoustic noise are generally produced at the point of phase commutation. Hence, these problems can be relieved by both the optimization of machine dimension [35] and the control of the commutation angle [36].

The phase current pulse conventionally has to synchronize with the rotor position of the SR machine, in order to produce the torque properly. However, the position sensors not only increase the construction cost and system complexity, but also decrease the reliability of the SR machine [37]. Upon the analysis of the variation of the flux linkages among the phases, the sensorless control algorithm can be achieved [38].

2.3.5   Comparisons of Traditional Magnetless Machines

The comparisons between the four basic magnetless brushless machines are given in Table 2.1. The comparisons are based on the key features, including the operating principle, rotor structure, control scheme, efficiency, torque range and speed range.

2.4   Upcoming Trends

Apart from the aforementioned accomplishments, new developments of the magnetless topologies have received well attentions recently.

2.4.1   Derived Structures from Permanent-Magnet Machines

With reference to the design philosophies from the profound PM machines, the advanced magnetless machines can then be produced. Therefore, the derived

Table 2.1 Comparisons of traditional magnetless brushless machines

	Induction	SynR	VR	SR
Operating principle	Induction	Reluctance	Reluctance	Reluctance
Rotor structure	Squirrel cage	Axially laminated	Salient pole	Salient pole
Control scheme	Rotating field	Rotating field	Rotating field	Switching pulse
Efficiency	High	Medium	Medium	Low
Torque range	Low	Medium	High	Low
Speed range	High	Medium	Low	High

magnetless machine can share the similar characteristics as well as advantages from its ancestors.

2.4.1.1 Doubly Salient DC-Field Machine

Shared the same design philosophy as the DSPM machine [7], as shown in Fig. 2.5, the doubly salient DC-Field (DSDC) machine can also incorporate the benefits from both the SR and the PM brushless machines. Moreover, the DSDC machine can utilize its independent DC-field excitation for flux-regulation, and hence its overall efficiency can be improved under different situations, namely at different loads and different operating speed ranges [39]. Inheriting the characteristics from its predecessor, the ideal back electromotive force (EMF) waveform can be classified as the trapezoidal [40] or sinusoidal patterns [41], depending on the machine structure.

Apart from the conventional three-phase 6/4-pole topology [42], as shown in Fig. 2.6, other slot/pole combinations, e.g., three-phase 12/8-pole [39], three-phase 24/32-pole [43] and five-phase 10/8-pole topologies [44] have also been developed. All developed machines exhibit different characteristics, in terms of power densities, torque densities, operating ranges, for various applications. Despite the profound symmetric stator/rotor pole combination, the asymmetric stator/rotor pole machines are also proposed [45]. Based on the new suggested arrangement, the DSDC machines with odd-number rotor-pole structure are developed. As a result, higher design flexibility can be provided.

Except the conventional single-stator single-rotor structure, the double-stator single-rotor topologies have also shown some attractive features. The most common

Fig. 2.5 Three-phase 6/4-pole doubly salient permanent-magnet machine [7]

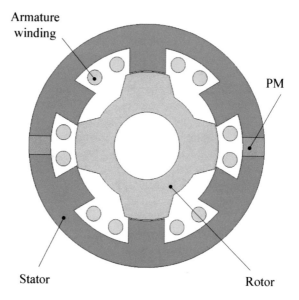

Armature winding

PM

Stator

Rotor

Fig. 2.6 Three-phase
6/4-pole doubly salient
DC-field machine [42]

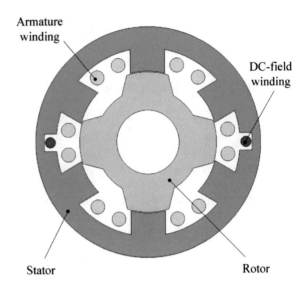

Armature winding

DC-field winding

Stator   Rotor

double-stator single-rotor DSDC machine employs the radial topology with the symmetrical structure for both the outer- and inner-segments [46]. With the double-stator topology, the inner spacing of the machine can be utilized, so that the power density can be improved. To provide special characteristics, the double-stator DSDC machine with asymmetrical structures between the outer- and the inner-segments, is also proposed [47]. Meanwhile, the double-stator structure towards its radial axis has also been suggested [48].

2.4.1.2 Flux-Switching DC-Field Machine

Same as the FSPM machines, as shown in Fig. 2.7, that have drawn most of the attentions among its competitive group [49], its derived magnetless candidate, namely the flux-switching DC-field (FSDC) machine has also become very popular recently. Inheriting the characteristics from its ancestor, the FSDC machine exhibits the bipolar flux-linkage patterns, so that higher power density can be produced [50]. To achieve the bipolar flux-linkage patterns, there are generally two fundamental DC-field winding arrangements available, namely the toroidal-field winding arrangement [51] and the wound-field winding arrangement [52], as shown in Figs. 2.8 and 2.9, respectively. Even though the toroidal-field winding machine can potentially produce higher torque density, it suffers from more severe saturation problem within its stator iron yoke. On the other hand, the magnetic field of the would-field winding machine is radially excited and not concentrated, so that the saturation problem can be relieved.

Unlike the SR machine, which produces torque by only half of the torque producing zone, the FSDC machine can instead utilize the whole region for torque

Fig. 2.7 Flux-switching permanent-magnet machine [49]

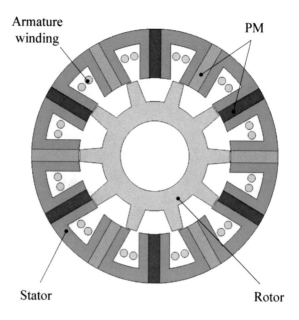

Fig. 2.8 Flux-switching DC-field machine with toroidal-field winding arrangement [51]

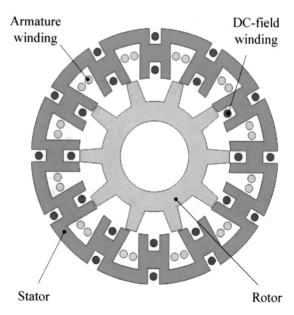

production. As compared with the profound magnetless SR machine, the FSDC machine not only enjoys the definite advantages of higher power and torque densities, but also the smaller radial force [53]. Therefore, the FSDC machine generally results less vibration and acoustic noise than the SR machine.

Fig. 2.9 Flux-switching
DC-field machine with
wound-field winding
arrangement [52]

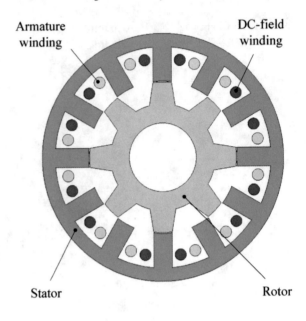

Armature
winding

DC-field
winding

Stator   Rotor

Fig. 2.10 Double-rotor
flux-switching DC-field
machine [54]

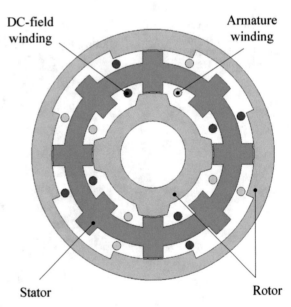

DC-field
winding

Armature
winding

Stator   Rotor

With the attempt to further utilize the inner spacing of the FSDC machine, the
double-rotor flux-switching DC-field (DR-FSDC) machine, has been proposed [54],
as shown in Fig. 2.10. Based on the toroidal winding arrangement, the armature and
DC-field windings can provide the symmetrical flux towards both rotors. With the

Table 2.2 Comparisons of advanced machines

	PM	DSDC	FSDC
Phase flux	Unipolar/Bipolar	Unipolar	Bipolar
Efficiency	High	Medium	High
Power density	Very high	Medium	High
Flux controllability	Low	Medium	High
Operating range	Narrow	Wide	Very wide
Manufacturing complexity	High	Low	Medium
Cost-effectiveness	Low	Medium	High

simple and robust structure, the developed DR-FSDC machine is very suitable for the rooftop application.

2.4.1.3   Comparisons of Advanced Machines

The comparisons between the two advanced magnetless brushless machines and the PM counterpart are given in Table 2.2. The comparisons are based on the key features, including the phase flux, efficiency, power density, flux controllability, operating range, manufacturing complexity and cost-effectiveness.

2.4.1.4   Technologies for Power and Torque Improvements

Without the installation of the high-energy-density PM materials, the magnetless machine suffers from the definite disadvantage of poor power and torque performances than the PM counterparts. In order to improve the market competitiveness of the magnetless machines, the technologies to improve the power and torque performances have been suggested.

2.4.1.5   Multi-Tooth Structure

To improve the torque density, the multi-tooth structure, which employs the multiple tooth per stator pole, has been developed [55], as shown in Fig. 2.11. Upon the multi-tooth topology, the magnetic flux within the stator yoke can be regulated in a way to transfer the energy to the rotor simultaneously, so that the torque density can be improved. In particular, the multi-tooth machine is favorable for the high-torque low-speed operation. In order word, the more the number of tooth per stator pole, the higher the torque density can be potentially achieved. However, the maximum number of the multiple tooth is limited by the physical constrain and the saturation within the tooth.

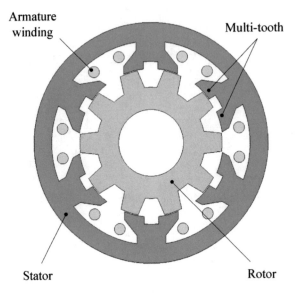

Fig. 2.11 Multiple tooth per stator pole structure [55]

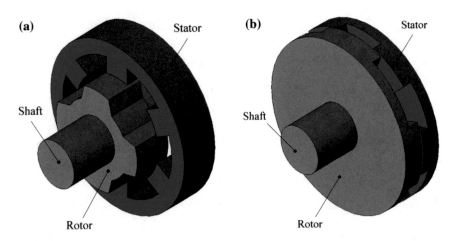

Fig. 2.12 Machine structures [56]. **a** Radial-field type. **b** Axial-field type

2.4.1.6 Axial-Field Structure

Despite the traditional radial-field (RF) structure, the magnetless machines can be developed based on the axial-field (AF) topology [56], as shown in Fig. 2.12. Upon the employment of the radial area for the torque production, the AF machines can produce higher power and torque densities as compared with its RF counterpart does [57]. Since the AF machines are derived from the RF topologies, theoretically every RF machine should consist of its corresponding AF derivation. As one of the most mature machine, the AF induction machine has been developed [58]. According to the simulation and experimental results, the AF induction machine can provide better performances, including higher efficiency and higher starting torque, than the RF counterparts do.

Same as the induction machine, another traditional magnetless machine, the SR machine has been further extended as the axial-field switched reluctance (AF-SR) machine. The sandwiched-stator double-sided-rotor structure with the toroidal winding arrangement has been proposed [59]. In particular, the AF-SR machine with higher rotor number is more preferable for the high-torque low-speed applications, such as the EVs or electric ship propulsions.

To reduce the iron losses within the yoke, the double-face printed circuit technology for the multilayer winding arrangement, has been implemented into the AF machine [60]. The torque of the double-face printed machine is produced by the interaction between the homopolar field and the rotating axial field created by the DC-field and three-phase armature winding, respectively. The double-face printed machine employs the solid steel discs as the rotor part, and therefore high robustness can be achieved for the flywheel applications.

Even though the AF topology enjoys some superior advantages, as compared with the well-developed RF counterpart, this type of machine suffers from the immature technology and manufacturing complexity.

2.4.1.7 High-Temperature Superconducting Materials

The power density achieved by the copper-field winding is relatively lower than that from the PM materials. To improve the situation, the high-temperature superconducting (HTS) materials have been implemented to the existing machine structures. There are two typical HTS implementations available, namely the HTS-field winding implementation [61] and the HTS bulk implementation [62].

With the advantages of high current density and thus high power density, the HTS-field winding implementation is suitable for the machine, which consists of limited winding slot area. In particular, this implementation has been employed to form the flux-switching high-temperature superconducting (FSHTS) machine [63].

Instead of improving the power density directly, the HTS bulk can be used to serve as the flux regulator to relieve the flux leakage problem. With the reduction of the flux leakage, the machine with the HTS bulk can therefore achieve higher power density. There is no conflict between the two HTS implementations, so that the hybrid HTS machine that incorporates the two implementations has been developed [64], as shown in Fig. 2.13. Despite the outstanding performances, the special cryogenic system is needed to keep the HTS materials to work properly.

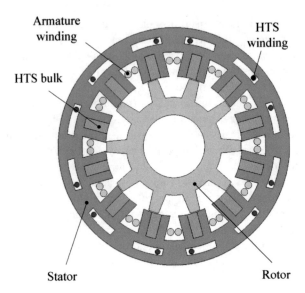

Fig. 2.13 High-temperature superconducting machine [64]

Table 2.3 Comparisons of performance improving technologies

	Multi-tooth	AF topology	HTS materials
Improved area	Torque density	Power and torque densities	Power and torque densities
Applicable machine type	Salient pole machine	All	Field excitation machine
Constrain	Not suitable for high-speed operation	Immature technology	Cryogenic issue
Developing rate	Medium	High	Medium

2.4.1.8 Comparisons of Performance Improving Technologies

The comparisons between the four upcoming technologies for performance improvements are given in Table 2.3. The comparisons are based on the key features, including the improved area, applicable machine type, constrain and developing rate.

2.4.2 Linear Topology

Due to the widespread of applications, the rotary machines have drawn most of the attentions in the past century. Yet, upon the enhancing attentions on the mass transit technology, there are increasing needs on the development of linear machines. The

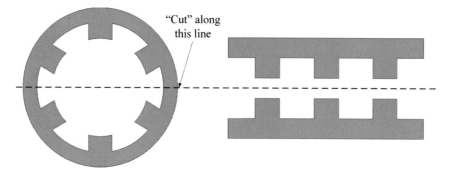

"Cut" along
this line

Fig. 2.14 Analogue between rotary and linear machines [65]

basic structure of the linear machine can be regarded as the extension from the rotary counterparts [65], as shown in Fig. 2.14.

Same as the AF machine, the linear machines can be derived from its corresponding rotary ancestors. As one of the most mature candidates, the linear SR machine has been developed based on the rotary SR machine [66]. The design considerations of the translator, as the analogue of the rotor in the rotary machine, are well studied. In addition to the linear SR machine, there are researches on the linear induction machine [67] and linear FSDC machine [68]. To improve the power density, the double-sided topology has been proposed [69]. Since the PM installations for the long-stator is very expensive, the magnetless linear machines have demonstrated very good potential for industrial applications.

2.4.3 Advanced Control Algorithm

As one of the most conventional control scheme, the direct-torque-control (DTC) has been widely employed for the brushless AC (BLAC) operating machines, while the traditional approach suffers from low estimation accuracy and system complexity. To improve the situation, the integration among the fifth-order filter, the high-pass filter and the logical calculation has been proposed [70]. Upon the adoption of the developed algorithm, better estimation capability and improved system quality can be achieved. In the meantime, the relationship between the independent field excitation of the DSDC machine and the system quality has been analyzed. Consequently, the space-vector pulse-width-modulation (SVPWM) has been implemented with the existing DTC scheme [71]. By utilizing the linear co-relation between the electromagnetic torque and the sine value of torque angle, the improved control scheme is formed, as shown in Fig. 2.15. Based on the improved DTC scheme, smaller torque pulsation and lower current total harmonic distortion (THD) are resulted.

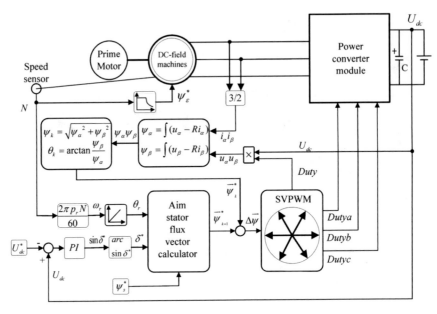

Fig. 2.15 Improved DTC scheme with the SVPWM [71]

2.5 Summary

The backgrounds, modern developments and upcoming technologies of magnetless brushless machines have been reviewed, with the clear descriptions on the operating principles, machine structures and control algorithms. The characteristics of the magnetless machines are discussed, with the highlights of the corresponding unique features. Upon the definite advantages of cost-effectiveness and flux regulation capability, the advanced magnetless machines have shown a great potential in some industry applications, such as EVs, wind power generations, ship propulsion systems and mass transit applications.

References

1. M. Cheng, W. Hua, J. Zhang, W. Zhao, Overview of stator-permanent magnet brushless machines. IEEE Trans. Ind. Electron. **58**(11), 5087–5101 (2011)
2. K.T. Chau, C.C. Chan, C. Liu, Overview of permanent-magnet brushless drives for electric and hybrid electric vehicles. IEEE Trans. Ind. Electron. **55**(6), 2246–2257 (2008)
3. A. Emadi, Y.J. Lee, K. Rajashekara, Power electronics and motor drives in electric, hybrid electric, and plug-in hybrid electric vehicles. IEEE Trans. Ind. Electron. **55**(6), 2237–2245 (2008)

4. D. Dorrell, L. Parsa, I. Boldea, Automotive electric motors, generators, and actuator drive systems with reduced or no permanent magnets and innovative design concepts. IEEE Trans. Ind. Electron. **61**(10), 5693–5695 (2014)
5. N. Tesla, A new system of alternate current motors and transformers. Trans. Am. Inst. Electr. Eng. **5**(V), 308–324 (1888)
6. S.E. Rauch, L.J. Johnson, Design principles of flux-switch alternators. Trans. Am. Inst. Electr. Eng. **74**(3), 1261–1268 (1955)
7. Y. Liao, F. Liang, T.A. Lipo, A novel permanent magnet motor with doubly salient structure. IEEE Trans. Ind. Appl. **31**(56), 1069–1078 (1995)
8. R.P. Deodhar, S. Andersson, I. Boldea, T.J.E. Miller, The flux-reversal machine: a new brushless doubly-salient permanent-magnet machine. IEEE Trans. Ind. Electron. **33**(4), 925–934 (1997)
9. E. Hoang, A.H. Ben-Ahmed, J. Lucidarme, Switching flux permanent magnet polyphased machines, in *Proceeding of Europe Conference Power Electronic Application*, Trondheim, Norway, pp. 903–908, September 1997
10. J.K. Kostko, Polyphase reaction synchronous motors. J. Am. Inst. Electr. Eng. **42**(11), 1162–1168 (1923)
11. C.H. Lee, Vernier motor and its design. IEEE Trans. Power Apparatus Syst. **82**(66), 343–349 (1963)
12. S.A. Nasar, D.C.-switched reluctance motor. Proc. IEEE **116**(6), 1048–1049 (1969)
13. P.L. Alger, R.E. Arnold, The history of induction motors in America. Proc. IEEE **64**(9), 1380–1383 (1976)
14. V.A. Fynn, Single-phase squirrel-cage motor with large starting torque and phase compensation. Proc. Am. Inst. Electr. Eng. **34**(10), 2215–2240 (1915)
15. G.R. Slemon, Modeling of induction machines for electric drives. IEEE Trans. Ind. Appl. **25**(6), 1126–1131 (1989)
16. E.A. Klingshirn, High phase order induction motors – Part I – Description and theoretical considerations. IEEE Trans. Power Apparatus Syst. **PAS-102**(1), 47–53 (1983)
17. J. Holtz, Sensorless control of induction motor drives. Proc. IEEE **90**(8), 1359–1394 (2002)
18. J. Holtz, Sensorless control of induction machines – with or without signal injection? IEEE Trans. Ind. Electron. **53**(1), 7–30 (2006)
19. P.J. Lawrenson, S.K. Gupta, Developments in performance and theory of segmental-rotor reluctance motors. Proc. Inst. Electr. Eng. **114**(5), 645–653 (1967)
20. A.J.O. Cruickshank, A.F. Anderson, R.W. Menzies, Theory and performance of reluctance motors with axially laminated anisotropic rotors. Proc. Inst. Electr. Eng. **1184**(7), 887–894 (1971)
21. T. Matsuo, T.A. Lipo, Rotor design optimization of synchronous reluctance machine. IEEE Trans. Energy Convers. **9**(2), 359–365 (1994)
22. M. Sanada, K. Hiramoto, S. Morimoto, Y. Takeda, Torque ripple improvement for synchronous reluctance motor using an asymmetric flux barrier arrangement. IEEE Trans. Ind. Appl. **40**(4), 1076–1082 (2004)
23. N. Bianchi, S. Bolognani, D. Bon, M.D. Pre, Torque harmonic compensation in a synchronous reluctance motor. IEEE Trans. Energy Convers. **23**(2), 466–473 (2008)
24. G. Qishan, E. Andresen, G. Chun, Airgap permeance of vernier-type, doubly-slotted magnetic structures. IET Electr. Power Appl. **135**(1), 17–21 (1988)
25. K.C. Mukherji, A. Tustin, Vernier reluctance motor. Proc. Inst. Electr. Eng. **121**(9), 965–974 (1974)
26. C. Shi, J. Qiu, R. Lin, A novel self-commutating low-speed reluctance motor for direct-drive applications. IEEE Trans. Ind. Appl. **43**(1), 57–65 (2007)
27. T. Lin, J. Qiu, C. Shi, Novel dual-excitation self-commutating low speed reluctance motor, in *The International Conference of Electrical Machines and Systems*, Tokyo, Japan, pp. 1–5, November 2009

28. T. Lin, J. Qiu, C. Shi, Dual-rotor self-commutating low speed reluctance motors, in *The International Conference of Electrical Machines and Systems*, Incheon, South Korea, pp. 1733–1738, October 2010
29. S. Taibi, A. Tounzi, F. Piriou, Study of a stator current excited vernier reluctance machine. IEEE Trans. Energy Convers. **21**(4), 823–831 (2006)
30. T.J.E. Miller, Optimal design of switched reluctance motors. IEEE Trans. Ind. Electron. **49**(1), 15–27 (2002)
31. R. Krishnan, R. Arumugam, J.F. Lindsay, Design procedure for switched-reluctance motors. IEEE Trans. Ind. Appl. **24**(3), 456–461 (1998)
32. A.V. Radun, C.A. Ferreira, E. Richter, Two-channel switched reluctance stator/generator results. IEEE Trans. Ind. Appl. **34**(5), 1026–1034 (1998)
33. D.A. Torrey, Switched reluctance generators and their control. IEEE Trans. Ind. Electron. **49** (1), 3–14 (2002)
34. P.C. Desai, M. Krishnamurthy, N. Schofield, A. Emadi, Novel switched reluctance machine configuration with higher number of rotor poles than stator poles: concept to implementation. IEEE Trans. Ind. Electr. **57**(2), 649–659 (2010)
35. A.V. Radun, Design considerations for the switched reluctance motor. IEEE Trans. Ind. Appl. **31**(5), 1079–1087 (1995)
36. I. Husain, Minimization of torque ripple in SRM drives. IEEE Trans. Ind. Electron. **49**(1), 28–39 (2002)
37. M. Ehsani, B. Fahimi, Elimination of position sensors in switched reluctance motor drives: State of the art and future trends. IEEE Trans. Ind. Electron. **49**(1), 40–47 (2002)
38. I. Husain, M. Ensani, Rotor position sensing in switched reluctance motor drives by measuring mutually induced voltage. IEEE Trans. Ind. Appl. **30**(3), 665–672 (1994)
39. Y. Fan, K.T. Chau, Design, modeling, and analysis of a brushless doubly fed doubly salient machine for electric vehicles. IEEE Trans. Ind. Appl. **44**(3), 727–734 (2008)
40. Y. Li, C.C. Mi, Doubly salient permanent-magnet machine with skewed rotor and six-state commutating mode. IEEE Trans. Magn. **43**(9), 3623–3629 (2007)
41. M. Cheng, K.T. Chau, C.C. Chan, Q. Sun, Control and operation of a new 8/6 doubly salient permanent magnet motor drive. IEEE Trans. Ind. Appl. **39**(5), 1363–1372 (2003)
42. K.T. Chau, M. Cheng, C.C. Chan, Nonlinear magnetic circuit analysis for a novel stator doubly fed doubly salient machine. IEEE Trans. Magn. **38**(5), 2382–2384 (2002)
43. Z. Zhang, Y. Yan, Y. Tao, A new topology of low speed doubly salient brushless DC generator for wind power generation. IEEE Trans. Magn. **48**(3), 1227–1233 (2012)
44. D. Zhu, X. Qiu, N. Zhou, Y. Yan, A novel five phase fault tolerant doubly salient electromagnetic generator for direct driven wind turbine, in *The International Conference of Electrical Machines and Systems,* Wuhan, China, pp. 2418–2422, October 2008
45. X. Liu, Z.Q. Zhu, Stator/rotor pole combinations and winding configurations of variable flux reluctance machines. IEEE Trans. Ind. Appl. **50**(6), 3675–3684 (2014)
46. Z. Zhang, Y. Zhou, Y. Yan, Feature investigation of a new dual-stator doubly salient brushless DC generator with and without rotor-yoke, in *The International Conference of Electrical Machines and Systems*, Busan, South Korea, pp. 1026–1031, October 2013
47. Y. Tao, Z. Zhang, Y. Yan, Z. Bo, Analysis of a new dual-stator doubly salient brushless DC generator, in *World Non-Grid-Connected Wind Power and Energy Conference*, Nanjing, China, pp. 1–4, November 2010
48. W. Dai, T. Xiu, H. Wang, Y. Yan, Control of a novel dual stator doubly salient aircraft engine starter-generator, in *The 37th IEEE Power Electronics Specialists Conference*, Jeju, South Korea, pp. 1–5, June 2006
49. Z.Q. Zhu, J.T. Chen, Advanced flux-switching permanent magnet brushless machines. IEEE Trans. Magn. **46**(6), 1447–1453 (2010)
50. C. Pollock, H. Pollock, R. Barron, J.R. Coles, D. Moule, A. Court, R. Sutton, Flux-switching motors for automotive applications. IEEE Trans. Ind. Appl. **42**(5), 1177–1184 (2006)
51. Y. Tang, J.J.H. Paulides, T.E. Motoasca, E.A. Lomonova, Flux-switching machine with DC excitation. IEEE Trans. Magn. **48**(11), 3583–3586 (2012)

52. A. Zulu, B.C. Mecrow, M. Armstrong, A wound-field three-phase flux-switching synchronous motor with all excitation sources on the stator. IEEE Trans. Ind. Appl. **46**(6), 2363–2371 (2010)
53. C. Pollock, M. Brackley, Comparison of the acoustic noise of a flux-switching and a switched reluctance drive. IEEE Trans. Ind. Appl. **39**(3), 826–834 (2003)
54. C. Yu, S. Niu, Development of a magnetless flux switching machine for rooftop wind power generation. IEEE Trans. Energy Convers. **30**(4), 1703–1711 (2015)
55. J. Faiz, J. Raddadi, J.W. Finch, Spice-based dynamic analysis of a switched reluctance motor with multiple teeth per stator pole. IEEE Trans. Magn. **38**(4), 1780–1788 (2002)
56. S.C. Oh, A. Emadi, Test and simulation of axial flux-motor characteristics for hybrid electric vehicles. IEEE Trans. Veh. Technol. **53**(3), 912–919 (2004)
57. F. Profumo, Z. Zhang, A. Tenconi, Axial flux machines drives: a new viable solution for electric cars. IEEE Trans. Ind. Electron. **44**(1), 39–45 (1997)
58. Z. Nasiri-Gheidari, H. Lesani, Investigation of characteristics of a single-phase axial flux induction motor using three-dimensional finite element method and d-q model. IET Electr. Power Appl. **7**(1), 47–57 (2013)
59. R. Madhavan, B.G. Fernandes, Performance improvement in the axial flux-segmented rotor-switched reluctance motor. IEEE Trans. Energy Convers. **29**(3), 641–651 (2014)
60. N. Bernard, H.B. Ahmed, B. Multon, Design and modeling of a slotless homopolar axial-field synchronous machine for a flywheel accumulator. IEEE Trans. Ind. Appl. **40**(3), 755–762 (2004)
61. C.H. Joshi, C.B. Prum, R.F. Schiferl, D.I. Driscoll, Demonstration of two synchronous motors using high temperature superconducting field coils. IEEE Trans. Appl. Supercond. **5**(23), 968–971 (1995)
62. F.N. Werfel, U. Flogel-Delor, D. Wippich, R. Rothfeld, A large scale approach of bulk HTS to the electric utility area. IEEE Trans. Appl. Supercond. **9**(2), 2018–2021 (1999)
63. Y. Wang, J. Sun, Z. Zou, Z. Wang, K.T. Chau, Design and analysis of a HTS flux-switching machine for wind energy conversion. IEEE Trans. Appl. Supercond. **23**(3), 5000904 (2013)
64. J. Rao, W. Xu, Modular stator high temperature superconducting flux-switching machines. IEEE Trans. Appl. Supercond. **24**(5), 0601405 (2014)
65. I. Boldea, S.A. Nasar, Linear electric actuators and generators. IEEE Trans. Energy Convers. **14**(3), 712–717 (1999)
66. B.S. Lee, H.K. Bae, P. Vijayraghavan, R. Krishnan, Design of a linear switched reluctance machine. IEEE Trans. Ind. Appl. **36**(6), 1571–1580 (2010)
67. A.S. Abdel-Khalik, S. Ahmed, A. Massoud, A five-phase linear induction machine with planar modular winding, in *The IEEE Conference on Industrial Technology*, Seville, Spain, pp. 580–585, March 2015
68. W. Li, K.T. Chau, C. Liu, C. Qiu, Design and analysis of a flux-controllable linear variable reluctance machine. IEEE Trans. Appl. Supercond. **24**(3), 5200604 (2014)
69. S.E. Abdollahi, M. Mirzayee, M. Mirsalim, Design and analysis of a double-sided linear induction motor for transportation. IEEE Trans. Magn. **51**(7), 8106307 (2015)
70. Y. Wang, Z. Deng, An integration algorithm for stator flux estimation of a direct-torque-controlled electrical excitation flux-switched generator. IEEE Trans. Energy Convers. **27**(2), 411–420 (2012)
71. Y. Wang, Z.Q. Deng, Analysis of electromagnetic performance and control schemes of electrical excitation flux-switching machine for DC power systems. IEEE Trans. Energy Convers. **27**(4), 844–855 (2012)

Part I
Modeling, Numerical Analysis and Experimental Verification of Torque Improving Topologies

Chapter 3
Multi-tooth Machines—Design and Analysis

3.1 Introduction

Because of the enhancing concerns on the energy utilization as well as the environmental protection, more attentions have been placed on the development of electric machines. In particular, the electric machines have to provide several distinctive characteristics, namely high efficiency, high power density, high controllability, wide speed range and maintenance-free operation [1]. The permanent-magnet (PM) machines, which can achieve most of the goals, have been actively developed [2]. Yet, due to the limited and fluctuating supply of the PM materials, the construction cost of the PM machines has been raised tremendously. Therefore, the magnetless machines, without any PM materials, have become more attractive recently [3].

The switched reluctance (SR) machine, as compared with the PM counterparts, enjoys the absolute merits of low cost, high robustness and excellent high-speed operation. However, similar as other magnetless machines, the SR machine also suffers from relatively poor torque density [4] and hence this type of machine is not suitable for high-torque low-speed operation. To relieve the situation, the multi-tooth SR (MSR) machine, which is particularly suitable for the high-torque low-speed operation, was proposed [5]. In the meantime, the concept of double-stator (DS) topology, which can drastically improve the torque density of different types of machines, was also investigated [6].

The purpose of this chapter is to firstly implement the concept of DS structure into the SR topology and the MSR topology to form the DS-SR machine and the DS-MSR machine, respectively. The design criteria and operating principles of three magnetless brushless machines, namely the SR, DS-SR and DS-MSR, will be described. To reduce the torque pulsation of the reluctance machine, a new design structure, so-called the mechanical-offset (MO), is also proposed. Based on the proposed MO design, the outer and inner rotor teeth of the DS machine are purposely mismatch by a conjugated angle, so that both of the stators can transfer

© Springer Nature Singapore Pte Ltd. 2018
C. H. T. Lee, *Design, Analysis and Application of Magnetless Doubly Salient Machines*, Springer Theses,
https://doi.org/10.1007/978-981-10-7077-8_3

powers and produce the complementary torques with each other. This mismatch structure can result with the offset performance to minimize the torque ripple. With the adaption of the finite element method (FEM), the entire proposed machine performances will be verified and quantitatively compared.

3.2   Magnetless Machines with Reluctance Topologies

The structures of the three proposed magnetless brushless machines, namely the 12/8-pole (twelve-stator-pole/eight-rotor-pole) SR machine, the 12/8-pole DS-SR machine and the 36/32-pole DS-MSR machine are shown in Fig. 3.1. Both DS topologies consists of 12 stator poles for both the outer and inner stators, while the MSR topology has 3 teeth per stator pole, and hence resulting with 36 equivalent stator poles.

The DS-SR machine is derived based on the implementation of DS structure and the SR machine, such that its pole arrangement can be described same as the SR machine equations as

$$\begin{cases} N_s = 2mk \\ N_r = N_s \pm 2k \end{cases} \tag{3.1}$$

where N_s is the number of stator poles, N_r the number of rotor poles, m the number of phases and k a positive integer.

Similarly, the DS-MSR machine is derived from the implementation of DS structure and the MSR machine, such that its design criteria can be expanded from that of the MSR machine as given by

$$\begin{cases} N_{sp} = 2mj \\ N_{se} = N_{sp}N_{st} \\ N_r = N_{se} \pm 2j \end{cases} \tag{3.2}$$

where N_{se} is the number of equivalent stator poles, N_{sp} the number of stator poles, N_{st} the number of stator teeth per pole and j a positive integer.

By selecting $k = 2$ and $m = 3$, it ends up with $N_s = 12$ and $N_r = 8$, and results the proposed structure for the SR and DS-SR machines. To achieve a fair comparison among all three magnetless machines, the number of conduction phases and winding configurations are purposely set to be the same. Hence, by selecting $j = 2$, $N_{sp} = 12$ and $N_{st} = 3$, it yields $N_{se} = 36$ and $N_r = 32$, and results the proposed structure for the DS-MSR machine. The DS-SR and DS-MSR machines, as compared with the SR machine, consist of higher manufacture complexity because of the adaption of double stators. Yet, if there are free from PM material, the corresponding construction is practicable. The key characteristics of all magnetless brushless machines are summarized as follows

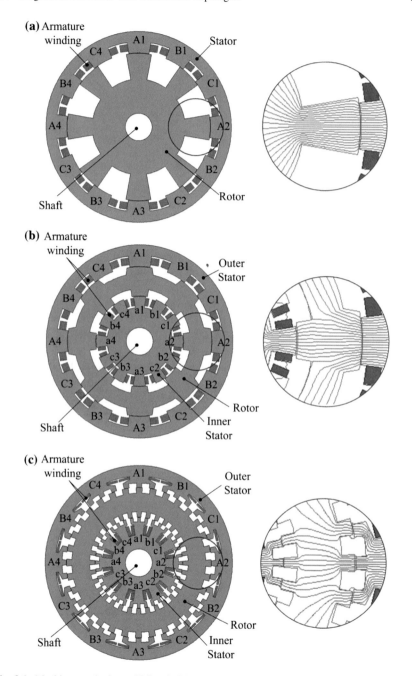

Fig. 3.1 Machine topologies. **a** 12/8-pole SR machine. **b** 12/8-pole DS-SR machine. **c** 36/32-pole DS-MSR machine

- Most of the inner rotor spaces of the SR machine are not utilized, and hence the torque density is degraded. Therefore, its torque performance is far lower than that of the PM machines.
- The inner spaces of the DS machines can be utilized for power transfer, and hence the overall torque density is improved.
- The MSR machine, which offers the flux-modulation effect, can offer higher torque density than its SR counterpart.
- The proposed reluctance machines all enjoy the absolute cost-benefit due to the magnetless structure.

3.3 Operating Principle

For the SR machine, two opposite coils pairs, namely the A1, A3, A2 and A4 coils are conducted in series, while the corresponding phases B and C perform the same conduction arrangement such that the symmetrical magnetic field is achieved with the magnetic polarities as N-S-N-S-N-S-N-S-N-S-N-S. For the DS machines, the inner coils, named as the lower case, consist of the same magnetic arrangements as the outer one, named as the capital case, such that both stators can transfer power to the rotor simultaneously.

All three machines adopt the same principle of operation and the theoretical waveforms are shown in Fig. 3.2, where i is the armature current that is applied during the period of increasing self-inductance L and T_e is the resulting electromagnetic torque. The torque equation can be described by

$$T_e = \frac{1}{2} i^2 \frac{dL}{d\theta}$$ (3.3)

Same to the SR machine, both DS-SR and DS-MSR machines employ the same speed control scheme, so that their operating speed can be governed by the value of N_r and the operating frequency as

$$n = \frac{60 f_{PH}}{N_r}$$ (3.4)

where n is the rotor speed and f_{PH} the commutating frequency. As illustrated by Eq. (3.4), the value of N_r of the DS-MSR machine is obviously larger than that of

Fig. 3.2 Theoretical operating waveforms

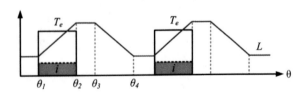

the SR and DS-SR machines. Hence, the DS-MSR machine can be operated at lower speed while producing relatively higher torque based on the same operating frequency. Because the core loss increases with the commutating frequency, the efficiency of the MSR machine is lower than its SR counterpart does during high-speed operation [5].

3.4 Comparisons of Switched Reluctance Machines

All the magnetless machines are compared based on the fair condition, namely the stack length, outer stator diameter, shaft diameter and airgap length are set equally. In addition, all the proposed machines are purposely designed to avoid magnetic saturation. Hence, the corresponding core losses are minimized so that the entire proposed machines can be compared fairly. By employing the FEM, all important machine performances of three proposed magnetless machines can be analyzed. Therefore, a quantitative comparison among them can be achieved. The key machine design data are shown in Table 3.1.

The airgap flux density distributions of the machines are calculated as shown in Fig. 3.3. The SR and DS-SR machines, as expected, share the same flux patterns because both of them employ the same machine topologies and principle of operations. In the meantime, the airgap flux density distributions of the DS-MSR machine are different from the others, where the corresponding flux of each stator pole is modulated as three portions with the number of teeth per each stator pole accordingly. It should also be noted the outer airgap flux density and the inner airgap flux density of both DS machines share the same pattern of waveforms. The results confirm that there is no distortion between the outer and inner airgap fluxes. In addition, the inner airgap flux density of the DS-MSR machine is slightly smaller than that of the outer airgap. This phenomenon can be explained by the fact that

Table 3.1 Key design data of the switched reluctance machines

Item	SR	DS-SR	DS-MSR
Rotor outside diameter (mm)	216.0	216.0	216.0
Rotor inside diameter (mm)	40.0	131.2	131.2
Outer stator outside diameter (mm)	280.0	280.0	280.0
Outer stator inside diameter (mm)	217.2	217.2	217.2
Inner stator outside diameter (mm)	N/A	130.0	130.0
Inner stator inside diameter (mm)	N/A	40.0	40.0
Outer and Inner airgap length (mm)	0.6	0.6	0.6
Stack length (mm)	80.0	80.0	80.0
No. of phases	3	3	3
No. of armature turns in outer stator	80	80	80
No. of armature turns in inner stator	N/A	80	80

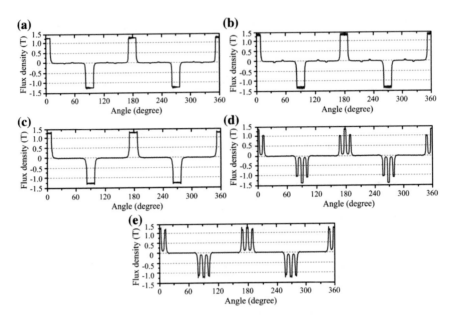

Fig. 3.3 Airgap flux distributions. **a** SR machine. **b** Outer airgap of DS-SR machine. **c** Inner airgap of DS-SR machine. **d** Outer airgap of DS-MSR machine. **e** Inner airgap of DS-MSR machine

the inner stator has higher possibility to have saturation, where relatively more severe situation is happened for the multi-tooth topology.

The static torque capabilities between the aligned and unaligned positions, as shown in Fig. 3.4, can be obtained by conducting the machines with different armature currents from 2 to 10 A with a step of 2 A. All proposed machines provide the torque performances align with the expectation given by Eq. (3.3). The results confirm that the machines do not experience severe magnetic saturation. In addition, the obtained torque comparisons verify the DS-MSR machine can achieve the highest torque under the same condition.

The steady output torque waveforms of the proposed machines at rated speeds are calculated as shown in Fig. 3.5. As illustrated, the average torques of the SR, DS-SR and DS-MSR machines are 19.1, 27.9 and 54.1 N m, respectively. The torque enhancement of the DS-SR and DS-MSR machines, as compared with the SR machine, can be up to 46.1 and 183.2%, respectively. The PM machines, with similar machine dimensions, can produce the rated torques ranging from 30 to 70 N m [2]. Hence, the proposed DS structure, even without installing any PM materials, can produce the torque density as comparable with the PM counterparts. Furthermore, the average torque values of the DS-SR machine when the inner stator is conducted alone, outer stator is conducted alone and both stators are conducted together are 9.6, 18.7 and 27.9 N m, respectively. The results confirm that the total torque produced by both stators is approximately equal to the summation of the

Fig. 3.4 Static torque
capabilities. **a** SR machine.
b DS-SR machine. **c** DS-MSR
machine

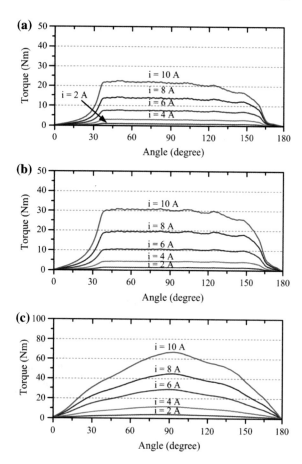

torque produced by the individual stators. Therefore, it suggests that both stators
can transfer power to the rotor simultaneously. Similar situation also applies for the
DS-MSR machine. In addition, the torque ripples of all machines are within the
acceptable range. Yet, because of the multi-tooth structure, the pulsating frequency
of the DS-MSR machine is larger than the other two machines.

The comparisons among the three magnetless machines, namely the SR
machine, the DS-SR machine and the DS-MSR machine are summarized in
Table 3.2. The results conclude that the DS-SR machine is favorable for low-torque
high-speed operation, while the DS-MSR machine is favorable for high-torque
low-speed operation.

Fig. 3.5 Steady torque
waveforms at rated speed.
a SR machine. **b** DS-SR
machine. **c** DS-MSR machine

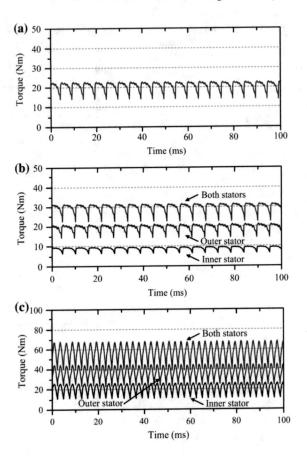

Table 3.2 Comparisons
among the switched
reluctance machines

Item	SR	DS-SR	DS-MSR
Power (W)	1000	1280	1360
Rated speed (rpm)	500	440	240
Outer airgap flux density (T)	1.29	1.41	1.38
Inner airgap flux density (T)	N/A	1.37	1.26
Rated torque (N m)	19.1	27.9	54.1
Torque enhancement	N/A	46.1%	183.2%

3.5   Proposed Structure for Torque Ripple Reduction

Undoubtedly, the MSR and DS structures can improve the torque densities of the
magnetless machines. However, the torque ripple issue, as another important criterion of the torque performance, has not been analyzed with details. To achieve
this goal, three additional MSR machines have been further developed.

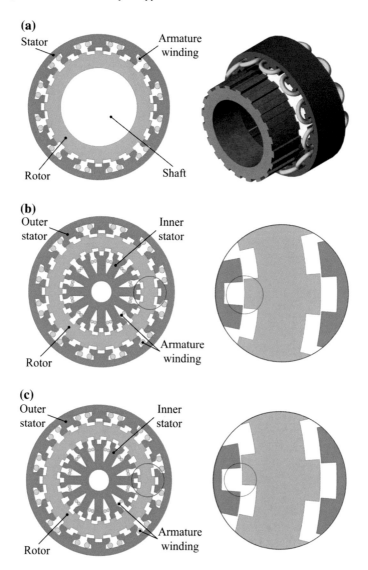

Fig. 3.6 Machine topologies. **a** MSR machine. **b** DS-MSR machine. **c** MO-DS-MSR machine

The structure of the conventional MSR machine, the DS-MSR machine and the proposed MO-DS-MSR machine are shown in Fig. 3.6.

Both the MSR and DS-MSR machines share the similar structure where both of them contain the outer stator of 12 salient poles, each fitted with 2 teeth and results with the equivalent stator teeth of 24. The corresponding stators are complied with the rotor of 20 poles. In the meantime, the SR machine consists of the single-stator single-rotor structure while the DS-MSR machine consists of the sided-stator sandwiched-rotor structure. It should be highlighted that the outer and the inner

Table 3.3 Key design data of the multi-tooth machines

Item	MSR	DS-MSR	MO-DS-MSR
Outer stator outside diameter (mm)	280.0	280.0	280.0
Outer stator inside diameter (mm)	217.2	217.2	217.2
Rotor outside diameter (mm)	216.0	216.0	216.0
Rotor inside diameter (mm)	151.2	151.2	151.2
Inner stator outside diameter	N/A	150.0 mm	150.0 mm
Inner stator inside diameter	N/A	40.0 mm	40.0 mm
Airgap length (mm)	0.6	0.6	0.6
Stack length (mm)	80.0	80.0	80.0
No. of stator poles	12	12	12
No. of stator teeth	2	2	2
No. of equivalent stator poles	24	24	24
No. of rotor poles	20	20	20
No. of armature phases	3	3	3
Mechanical-offset (°)	0	0	60
No. of turns per outer coil	50	50	50
No. of turns per inner coil	N/A	60	60

rotor teeth of the DS-MSR are aligned with each other, same as the conventional DS machines do. Therefore, to ease the control algorithm, its outer and inner armature windings are purposely connected in series.

Despite the proposed MO-DS-MSR machine share the similar topology as the DS-MSR machine does, its outer and inner rotor teeth are purposely mismatch with a conjugated angle to each other, as indicated in the circle. Hence, to transfer energy to the rotor properly, the outer and the inner windings should be controlled and conducted independently.

On the employment of the Eq. (3.2), all the MSR machines can be developed. By selecting $m = 3$, $j = 2$ and $N_{st} = 2$, this ends up with $N_{se} = 24$, $N_r = 20$, as the proposed machine topologies. All machine dimensions are purposely optimized to avoid the magnetic saturations as well as the core losses. The key machine design data is shown in Table 3.3. With the MO structure, the outer and inner torque components of the MO-DS-MSR machine are purposely mismatched with a conjugated angle to minimize the torque pulsation.

3.6 Proposed Algorithm for Torque Ripple Reduction

3.6.1 Conventional Conduction Algorithm

As suggested in earlier section, to drive the reluctance machines, a unipolar rectangular current I_{rect} is fed to the armature winding during the increasing period of its self-inductance L, so that $\theta_c = \theta_2 - \theta_1$, as shown in Fig. 3.7a. Under this conduction scheme, the corresponding reluctance torque can be described as by Eq. (3.3). As suggested, the average rated torque is controlled by the armature current value as well as the machine design. Despite the pulsating value is inversely proportional to the average rated torque, the increase of the armature current may also enlarge the magnitude of the torque ripple. Hence, the effect of the increased

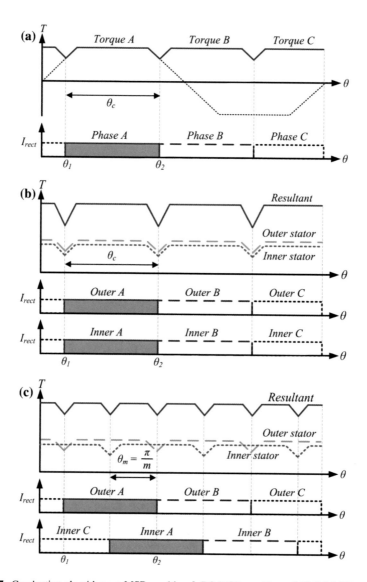

Fig. 3.7 Conduction algorithms. **a** MSR machine. **b** DS-MSR machine. **c** MO-DS-MSR machine

rated torque is unfavorably offset. Therefore, there is no improvement regarding to the torque pulsating problem, based on the increased armature currents.

The two stators of the conventional DS machine are operated simultaneously with no phase shift, as shown in Fig. 3.7b, so that each armature phase can be controlled as

$$\begin{cases} i_k = I_{rect} & \theta_1 \le \theta \le \theta_2 \\ i_k = 0 & 0 \le \theta \le \theta_1, \ \theta_2 \le \theta \le 2\pi \end{cases} \quad \text{for } k = 1, 2 \qquad (3.5)$$

where i_1 is the outer armature current and i_2 the inner armature current. With this conduction scheme, the two stators can transfer the power to the rotor simultaneously to boost up its torque density. Yet, the local maxima and local minima of the two torque components are unfavorably integrated with each other. Therefore, the DS machine with the conventional design can only increase its torque density, while the torque ripple problem has not been solved yet.

3.6.2   Proposed Mechanical-Offset Algorithm

Referring to the Eq. (3.3), the average torque magnitude is controlled by the relative position between the stator and the rotor teeth. Therefore, the local maxima and the local minima are also controlled by the relative stator-rotor position. Based on the proposed MO design, the outer and inner rotor teeth of the MO-DS-MSR machine are purposely offset with a conjugated electrical angle of $\theta_m = \pi/m$, as shown in Fig. 3.7c. Therefore, the two stators should be operated independently, where either one armature set employs the conventional conduction scheme as Eq. (3.5), while the other set employs the scheme as follows

$$\begin{cases} i_{1 \ or \ 2} = I_{rect} & \theta_1 + \theta_m \le \theta \le \theta_2 + \theta_m \\ i_{1 \ or \ 2} = 0 & \theta_m \le \theta \le \theta_1 + \theta_m, \ \theta_2 + \theta_m \le \theta \le 2\pi + \theta_m \end{cases} \qquad (3.6)$$

Based on the special offset configuration, the local maxima and the local minima are favorably mismatched with each other. Therefore, the torque pulsation produced by the outer and inner torque components can be compensated to generate the smoother resultant torque. To offer the better performances, the magnitudes and patterns of the two torque components should be adjusted as similar as possible. In the meantime, this task can be achieved by the optimization of the machine topology, the armature current magnitude and the winding arrangement.

Because the two torque components of the MO-DS-MSR machine are integrated with each other, the average value should not be deteriorated, as compared with the conventional DS machines do. Hence, the proposed MO-DS-MSR machine is expected to provide the same torque level as those from the conventional counterparts do. It should be noted the local maxima and local minima are spread in accordance to the conjugated positions, and hence the resultant torque with the MO design should result with higher pulsating frequency. In the meantime, the armature currents of the MO-DS-MSR machine should be the same as the conventional one. Hence, the copper loss produced by the MO machine should be the same as compared with its conventional counterpart.

3.7   Comparisons of Torque Ripple Performances

With the employment of the FEM analysis, the machine performances of the proposed machines can be deduced. The average torque and the corresponding pulsating values of the MSR, the DS-MSR and the MO-DS-MSR machines are shown in Fig. 3.8. As illustrated, the average torques of all three machines increase in accordance to the increased armature currents, and hence complying with the expectation of the Eq. (3.3). As suggested in the previous section, the improved torque density of the DS-MSR machine, as compared with the MSR machine, can provide very little improvement towards the torque pulsating problems. On the

Fig. 3.8 Average torque and torque ripple performances. **a** MSR machine. **b** DS-MSR machine. **c** MO-DS-MSR machine

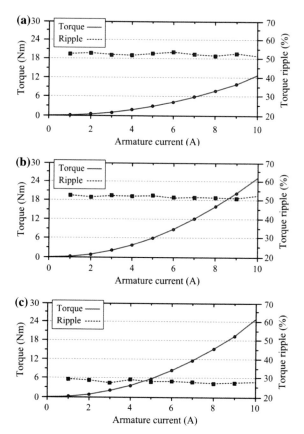

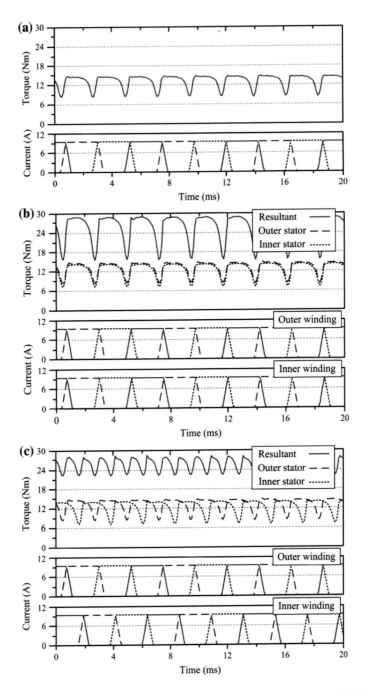

Fig. 3.9 Rated torque waveforms. **a** MSR machine. **b** DS-MSR machine. **c** MO-DS-MSR machine

Table 3.4 Comparisons among the multi-tooth machines

Item	MSR	DS-MSR	MO-DS-MSR
Power (W)	620	1220	1210
Rated speed (rpm)	450	450	450
Outer stator torque (N m)	N/A	13.2	13.2
Inner stator torque (N m)	N/A	12.7	12.7
Resultant torque (N m)	13.2	25.9	25.7
Torque ripple	51.6%	52.4%	27.9%

other hand, the MO-DS-MSR machine, which employs the proposed MO topology, can greatly minimize its torque pulsation value as compared with its counterparts.

The steady torque waveforms of the MSR, the DS-MSR and the MO-DS-MSR machines are shown in Fig. 3.9. When the armature excitation is set to be 10 A, the average steady torques of the MSR, the DS-MSR and the MO-DS-MSR machines are 13.2, 25.9, and 25.7 N m, respectively. The results verify the two torque components of the MO-DS-MSR can implement with each other seamlessly, so that it can produce the same torque level as compared with the conventional design.

With the adaption of the conventional design, the local maxima and the local minima of the outer and inner torque pulsations are unfavorably added up, as shown in Fig. 3.9b. This offsets the effect from the increased average torque so that it produces the torque pulsation of 52.4%. This particular value is similar to those produced from the MSR machine where the pulsation is around 51.6%, as shown in Fig. 3.9a. Therefore, the DS-MSR machine enjoys no advantage regarding the torque pulsation problem. In the meantime, based on the proposed MO design, the local maxima and local minima are purposely offset and favorably integrated with each other, as shown in Fig. 3.9c. Thus, the MO-DS-MSR machine can gain the entire advantages from the increased torque density to produce an improved pulsating torque value of 27.9%. It should also be noted that even under the same operating speed, the MO-DS-MSR machine, as compared with the other two machines, produces an increased torque pulsating frequency. The comparison results are summarized and tabulated in Table 3.4.

3.8 Summary

In this chapter, the performances of three magnetless brushless machines, namely the SR, DS-SR and DS-MSR machines, are discussed and thoroughly compared. The comparisons conclude that the DS-SR machine is favorable for low-torque high-speed operation, while the DS-MSR machine is favorable for high-torque low-speed operation. To be specific, the DS-SR machine and DS-MSR machine can offer better torque values than the SR machine by 46.1 and 183.2%, respectively. Because of the elimination of expensive PM materials, the magnetless DS machines take the absolute advantage of better cost-effectiveness than the PM machines do.

In the meantime, the torque pulsating performances of the reluctance machines have been discussed in details. The MO structure, which can purposely offset and favorably integrate the torque pulsation, is proposed. With the proposed MO topology, the DS machine can gain all the merits from the improved torque characteristics.

References

1. M. Cheng, W. Hua, J. Zhang, W. Zhao, Overview of stator-permanent magnet brushless machines. IEEE Trans. Ind. Electron. **58**(11), 5087–5101 (2011)
2. K.T. Chau, C.C. Chan, C. Liu, Overview of permanent-magnet brushless drives for electric and hybrid electric vehicles. IEEE Trans. Ind. Electron. **55**(6), 2246–2257 (2008)
3. C.H.T. Lee, K.T. Chau, C. Liu, C.C. Chan, Overview of magnetless brushless machines. IET Electric Power Appl. (in press)
4. D. Dorrell, L. Parsa, I. Boldea, Automotive electric motors, generators, and actuator drive systems with reduced or no permanent magnets and innovative design concepts. IEEE Trans. Ind. Electron. **61**(10), 5693–5695 (2014)
5. J. Faiz, J. Raddadi, J.W. Finch, Spice-based dynamic analysis of a switched reluctance motor with multiple teeth per stator pole. IEEE Trans. Magn. **38**(4), 1780–1788 (2002)
6. S. Niu, K.T. Chau, J.Z. Jiang, C. Liu, Design and control of a new double-stator cup-rotor permanent-magnet machine for wind power generation. IEEE Trans. Magn. **43**(6), 2501–2503 (2007)

Chapter 4
Double-Rotor Machines—Design and Analysis

4.1 Introduction

Because of the enhancing needs on the environmental protection and energy utilization, the development of electric vehicles (EVs) and renewable energy (RE) are speeding up in the past few decades. To improve the market penetration of these applications, as suggested in Chap. 3, the corresponding electric machines have to offer great performances, including high efficiency, high power density, high controllability, wide speed range and maintenance-free operation [1]. Due to the relatively simple and effective construction, the single-rotor permanent-magnet (PM) machine has been dominating the domestic markets [2]. However, the increased PM material prices have severely hindered the further development of PM machines. Yet, the magnetless brushless machines for EV and RE applications have attracted more attention recently [3].

There are increasing numbers of applications for two rotating loads, so that the double-rotor (DR) machine has started to attract more attentions recently. For instances, the DR machines consisting of two separated rotors can be directly connected to the wheels of the EVs to offer the electronic differential characteristics [4]. Furthermore, the separated rotors can be used to perform the power splitting for the internal combustion engine such that the hybrid EVs can enjoy the merits of better energy utilization [5]. The flux-modulated permanent-magnet (FMPM) machine, which employs the magnetic-gearing effect, has also adapted the DR topology to achieve high-torque low-speed characteristics [6]. In addition, the DR generator with the contra-rotating characteristic has been proposed for wind power generation recently [7].

The magnetless brushless machines, as compared with the PM counterparts, generally suffer from relatively lower torque density. To relieve the problem, as suggested, the multi-tooth switched reluctance (MSR) machine, which is favorable for high-torque low-speed operation, is one of the most promising solutions [8]. On the utilization of the DC-field excitation, the DC-field multi-tooth switched

© Springer Nature Singapore Pte Ltd. 2018
C. H. T. Lee, *Design, Analysis and Application of Magnetless Doubly Salient Machines*, Springer Theses,
https://doi.org/10.1007/978-981-10-7077-8_4

reluctance (DC-MSR) machine can provide the dual-mode operation, to improve the fault-tolerability [9]. In the meantime, by choosing a definite combination of stator and rotor pole numbers with the suitable DC-field excitation configurations, the flux-switching DC-field (FSDC) can be realized [10]. Same as the flux-switching permanent-magnet (FSPM) machine, the FSDC machine can provide the bipolar flux-linkage characteristics, and hence higher power density is achieved [11, 12].

In this chapter, by implementing the DC-MSR and FSDC machines with the DR topology, the DR-DC-MSR and DR-FSDC machines are newly proposed. These proposed machines are purposely designed for special direct-drive applications with two separated motoring wheels for EVs or with two sets of separately driven wind blades for wind power harvesting. Based on the employment of the independent DC-field excitation, the proposed machines can operate with two operating modes, namely the doubly salient DC (DSDC) mode and the MSR mode. In the meantime, the air-gap flux density of the proposed machines can be effectively controlled to achieve better energy efficiency. By the finite element method (FEM), the machine performances will be analyzed, and hence the feasibility for direct-drive applications with two separated rotating bodies can be verified.

4.2 Magnetless Machines with Double-Rotor Topology

With the employment of the DR topology, the inner spaces of the two proposed machines, namely the DR-DC-MSR machine and DR-FSDC machine, are purposely utilized to serve as the inner rotor, and hence the torque density can be improved. With the PM-free structure, the proposed magnetless brushless machines enjoy the absolute advantages of higher cost-effectiveness.

4.2.1 Double-Rotor DC-Field Multi-Tooth Switched Reluctance Machine

The proposed DR-DC-MSR machine, which contains the double-rotor sandwiched-stator structure, is shown in Fig. 4.1a. Based on the multi-tooth stator pole topology, the proposed machine can accomplish the flux-modulation effect to improve its torque performance. The outer machine segment typically employs a larger number of rotor poles because it consists of a larger circumferential cross-sectional area. On the other hand, the inner machine segment instead employs a smaller number of rotor poles. To comply this rationale, the outer stator contains 6 salient poles, each having 4 teeth, and hence serving as an equivalence of 24 stator teeth. This results with the outer rotor of 22 salient poles. In the meantime, the inner stator contains 6 salient poles, each having 2 teeth, and hence serving as an equivalence of 12 stator teeth. This results with the inner rotor of 10 salient poles.

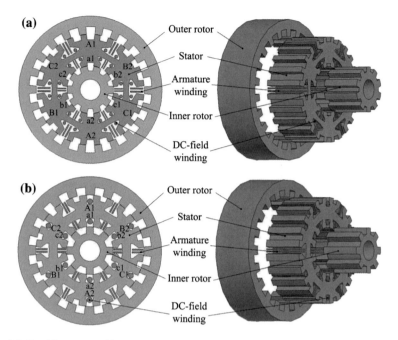

Fig. 4.1 Double-rotor machines. **a** DR-DC-MSR topology. **b** DR-FSDC topology

The proposed DR-DC-MSR machine installs with two independent sets of windings, namely the armature winding and the DC-field winding. Both windings employ the concentrated winding configuration where both of them are equipped with their magnetic axes in parallel to each other. Based on this winding config-uration, the DC flux-linkages towards the two rotors travel along the same direction, as shown in Fig. 4.2. The operating speeds of the outer rotor and the inner rotor can be controlled separately when the outer winding and inner armature winding are independently regulated. The two separated rotors can be operated simultaneously in the same direction if the armature windings are connected in series. Furthermore, the polarity of the DC flux-linkages remain the same even the rotors are rotated to different positions, such as the Position 1 and Position 2, as shown in Fig. 4.2.

The pole-pair arrangement of the proposed DR-DS-MSR machine can be described by the equations as follows

$$\begin{cases} N_{sp} = 2mj \\ N_{se} = N_{sp}N_{st} \\ N_r = N_{se} \pm 2j \end{cases} \quad (4.1)$$

where N_{sp} is the number of stator poles, N_{st} the number of stator teeth, N_{se} the number of equivalent stator poles, N_r the number of rotor poles, m the number of armature phases and j is any integer. By selecting $N_{sp} = 6$, $N_{st} = 4$ and $j = 1$, this results with $N_{se} = 24$ and $N_r = 22$ as the topology of the outer segment. In the

Fig. 4.2 DC flux-linkage
paths of DR-DC-MSR
machine. **a** Position 1.
b Position 2

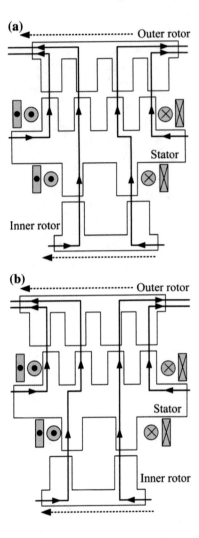

meantime, by selecting $N_{sp} = 6$, $N_{st} = 2$ and $j = 1$, this results up with $N_{se} = 12$ and $N_r = 10$ as the topology of the inner segment. The combinations end up with the proposed topology of the DR-DC-MSR machine.

4.2.2 Double-Rotor Flux-Switching DC-Field Machine

The structure of the DR-FSDC machine, which consists of a similar design process as the DR-DS-MSR machine does, is shown in Fig. 4.1b. The outer stator contains 6 salient poles, each having 4 teeth, and hence serving as an equivalence of 24 stator teeth. This results with the outer rotor of 20 salient poles. In the meantime, the inner stator consists of 6 salient poles, each having 2 teeth, and hence serving as an equivalence of 12 stator teeth. This results with the inner rotor of 8 salient poles.

The DR-FSDC machine employs the similar structure as the DR-DC-MSR machine, where both of them equip with two winding types. However, unlike the DR-DC-MSR machine that employs the concentrated winding configuration for both windings, the DR-FSDC machine employs the concentrated winding config-uration and the toroidal winding configuration for the armature winding and the DC-field winding, respectively. Based on these winding arrangements, the two winding sets are equipped with their magnetic axes perpendicular to each other. Because the two winding sets are placed within separated slots, more armature coils can be allocated for the DR-FSDC machine as compared with its counterpart do.

Fig. 4.3 DC flux-linkage paths of DR-FSDC machine. **a** Position 1. **b** Position 2

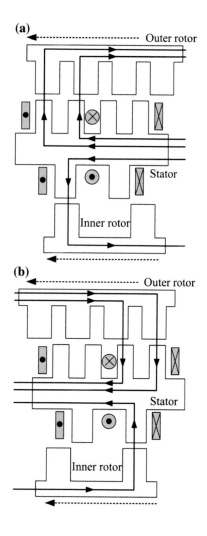

Furthermore, the two winding sets are physically decoupled so that the DR-FSDC machine can enjoy higher fault tolerability to the inter-turn short-circuited fault than the DR-DC-MSR machine, where both winding sets are hold up together. Based on the toroidal winding configuration, the DC flux-linkages of two rotors travel in opposite directions as shown in Fig. 4.3. In the meantime, when the armature windings are separated regulated, the two rotors can then be operated in the same direction. Furthermore, the polarity of the DC flux-linkage reverses, during the travelling of the rotor poles, as illustrated in Position 1 and Position 2, as shown in Fig. 4.3. Therefore, the flux-switching characteristic is realized such that better torque density can be achieved.

The pole-pair arrangement of this DR-FSDC machine can be described based on the equations as follows

$$\begin{cases} N_{sp} = 2mj \\ N_{se} = N_{sp}N_{st} \\ N_r = N_{se} - N_{sp} \pm 2j \end{cases} \qquad (4.2)$$

By selecting $N_{sp} = 6$, $N_{st} = 4$ and $j = 1$, this results with $N_{se} = 24$ and $N_r = 20$ as the topology for the outer segment. In the meantime, by selecting $N_{sp} = 6$, $N_{st} = 2$ and $j = 1$, this results with $N_{se} = 12$ and $N_r = 8$ as the topology for the inner segment. The combinations end up with the proposed topology for the DR-FSDC machine.

Table 4.1 Key data of the proposed double-rotor machines

Item	DR-DC-MSR	DR-FSDC
Outer rotor outside diameter (mm)	280.0	280.0
Outer rotor inside diameter (mm)	211.2	211.2
Stator outside diameter (mm)	210.0	210.0
Stator inside diameter (mm)	91.2	91.2
Inner rotor outside diameter (mm)	90.0	90.0
Inner rotor inside diameter (mm)	40.0	40.0
Airgap length of both segments (mm)	0.6	0.6
Stack length (mm)	80.0	80.0
No. of stator poles of both segments	6	6
No. of stator teeth of outer segment	4	4
No. of stator teeth of inner segment	2	2
No. of equivalent stator poles of outer segment	24	24
No. of equivalent stator poles of inner segment	12	12
No. of rotor poles of outer segment	22	20
No. of rotor poles of inner segment	10	8
No. of armature phases	3	3
No. of turns per outer armature coil	50	65
No. of turns per inner armature coil	40	55

To offer a fair comparing environment for the two proposed DR machines, the critical parameters, namely the stator diameter, rotor diameter, airgap length, stack length, number of equivalent stator poles and number of armature phases are set equal. The key design data are summarized and tabulated in Table 4.1.

4.3   Principle of Operation

Based on the adaption of the independent DC-field excitation, the proposed DR machines can be operated with two separated operating principles, namely the DSDC mode and the MSR mode. In particular, the DSDC mode and the MSR mode mainly serve for normal operation and fault-tolerant operation, respectively. Regarding the EV applications, the fault-tolerant operation is very important. For example, when the electric machine is abruptly malfunction under fault, the faulty EV may result severe traffic jam or even deathly accident. With the fault-tolerant operation, namely the MSR mode, the proposed DR machines can allow the EV at least to continue the minimum operation even under the DC-field winding fault.

4.3.1   Doubly Salient DC-Field Mode

In the case if the DC-field winding is function normally, two proposed DR machines can be operated by using the bipolar conduction algorithm, which is similar to the conventional operation employing on the doubly salient permanent-magnet (DSPM) machines [13, 14]. When the DC flux-linkage Ψ_{DSDC} is increasing and the no-load electromotive force (EMF) is positive, a positive armature current I_{BLDC} is injected to generate a positive torque T_{DSDC}. On the other hand, a negative armature current $-I_{BLDC}$ is injected when the Ψ_{DSDC} is decreasing and the no-load EMF is negative, and hence a positive torque is also generated. This operation mode is so-called as the DSDC mode and the theoretical waveforms are shown in Fig. 4.4a. Each phase undergoes 120° conduction with $\theta_2 - \theta_1 = \theta_4 - \theta_3 = 120°$. The generated electromagnetic torque T_{DSDC} can be described as

$$T_{DSDC} = \frac{1}{2\pi} \int_0^{2\pi} \left(i_{BLDC} \frac{d\psi_{DSDC}}{d\theta} + \frac{1}{2} i_{BLDC}^2 \frac{dL_D}{d\theta} \right) d\theta \tag{4.3}$$

where L_D is the self-inductance. Based on the DSDC mode, the torque is mainly generated by the DC-field torque component, while the reluctance torque component is the pulsating one with zero average value. Therefore, the pulsating reluctance torque component can be eliminated so that the torque expression can be further elaborated as

Fig. 4.4 Principle of
operations. **a** DSDC mode.
b MSR mode

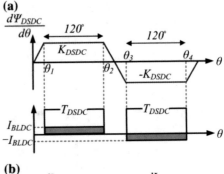

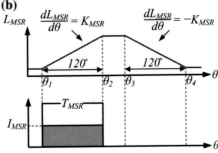

$$T_{DSDC} = \frac{1}{2\pi}\left(\int_{\theta_1}^{\theta_2} I_{BLDC}K_{DSDC}d\theta + \int_{\theta_3}^{\theta_4} (-I_{BLDC})(-K_{DSDC})d\theta\right) = \frac{2}{3}I_{BLDC}K_{DSDC}$$

(4.4)

where K_{DSDC} is the slope of Ψ_{DSDC} with respect to θ.

Same as the conventional DSPM machines, both proposed DR machines employ the same operating speed control scheme where the operating speed is controlled by the value of N_r and the operating frequency as

$$\omega = \frac{60f_{PH}}{N_r}$$

(4.5)

where ω is the rotor speed and f_{PH} the commutating frequency. The value of N_r of the outer segment of the proposed machines obviously is larger than those of the corresponding inner segment. Hence, even under the same operating frequency, the outer rotors consist of lower operating speeds with higher torques, as compared with the inner rotors.

4.3.2 Multi-tooth Switched Reluctance Mode

In the case if the DC-field current is under an open-circuit fault or short-circuit fault, the DC-field excitation can be terminated so that the proposed DR machines can then be operated by the unipolar conduction algorithm. To be specific, a unipolar rectangular current I_{MSR} is injected to the armature winding when the self-inductance L_{MSR} is increasing such that the reluctance torque T_{MSR} is positive within $\theta_2-\theta_1 = 120°$, as shown in Fig. 4.4b. This mode is so-called as the MSR mode. Even though the MSR mode can provide the basic functions of the DR machines, only half of the torque producing zone is used. Therefore, the torque performances are degraded such that the torque pulsation problem is more severe than that generated at the DSDC mode. Hence, the MSR mode should be employed as fault-tolerant operation for the occasions when the DC-field winding is under fault condition [15]. The generated reluctance torque at the MSR mode can be described as

$$T_{MSR} = \frac{1}{2\pi} \int_0^{2\pi} \left(\frac{1}{2} i_{MSR}^2 \frac{dL_{MSR}}{d\theta} \right) d\theta = \frac{1}{2\pi} \int_{\theta_1}^{\theta_2} \left(\frac{1}{2} I_{MSR}^2 K_{MSR} \right) d\theta = \frac{1}{6} I_{MSR}^2 K_{MSR} \quad (4.6)$$

where K_{MSR} is the slope of L_{MSR} with respect to θ. To keep the same torque magnitude between two modes, the Eqs. (4.4) and (4.6) that governs the torques from the DSDC mode and MSR mode should be equated as

$$I_{MSR} = 2\sqrt{\frac{I_{BLDC} K_{DSDC}}{K_{MSR}}} \quad (4.7)$$

According to Eq. (4.7), the proposed DR machines can provide the same average torque between two modes with the deduced armature current at the MSR mode. Nevertheless, the armature current at the MSR mode is typically larger than that at the DSDC mode. With the consideration of the pulsation issue and armature value, the corresponding torque performances and efficiencies at MSR mode are expected to be worse than that at the DSDC mode.

4.4 Electromagnetic Field Analysis

Electromagnetic field analysis has been employed for the development of the electric machines, where it can be typically classified as two major types, namely the analytical field calculation [16] and the numerical field calculation [17]. In this chapter, the analysis of the proposed DR machine performances is performed by the FEM. To develop the machine modeling, three equations are formed. Firstly, the electromagnetic field equation is described as [18, 19].

$$\begin{cases} \Omega : \frac{\partial}{\partial x}\left(v\frac{\partial A}{\partial x}\right) + \frac{\partial}{\partial y}\left(v\frac{\partial A}{\partial y}\right) = -(J_z + J_f) \\ A|_S = 0 \end{cases} \tag{4.8}$$

where Ω is the field solution region, A the z-direction components of vector potential, J_z the z-direction components of current density, J_f the equivalent current density of the excitation field, S the Dirichlet boundary and v the reluctivity. Secondly, the armature circuit equation of the machine under motoring is described as

$$u = Ri + L_e\frac{di}{dt} + \frac{l}{s}\iint\limits_{\Omega_e} \frac{\partial A}{\partial t}d\Omega \tag{4.9}$$

where u is the applied voltage, R the winding resistance, L_e the end winding inductance, l the axial length, s the conductor area of each turn of phase winding and Ω_e the total cross-sectional area of conductors of each phase winding. On the other hand, the circuit equation under generation is described as

$$u = \frac{l}{s}\iint\limits_{\Omega_e} \frac{\partial A}{\partial t}d\Omega - Ri - L_e\frac{di}{dt} \tag{4.10}$$

Thirdly, the motion equation of the machine is described as

$$J_m\frac{\partial \omega}{\partial t} = T_e - T_L - \lambda\omega \tag{4.11}$$

where J_m is the moment of inertia, T_L the load torque and λ the damping coefficient. These three set of equations can be employed to estimate the steady-state and transient machine performances of the proposed DRC machines. JMAG-Designer is employed as the magnetic solver to perform the finite element analysis for the proposed DR machines. The generated meshes and the corresponding no-load magnetic field distributions are shown in Figs. 4.5 and 4.6, respectively. A few hours are normally needed for a single simulation process based on a standard PC.

The no-load EMF waveforms of the proposed DR machines at the rated speed with the DC-field excitation as variation are shown in Fig. 4.7. To compare the two machines fairly, both of them are operated under the same operating frequency as 200 Hz. Both machines achieve similar characteristics where their no-load EMF increases linearly based on the DC-field excitation before 700 A-turn, just when the magnetic saturations start to happen. These results verify that the proposed machines can effectively regulate the flux densities by controlling the DC-field excitation to optimize the efficiency. In the meantime, because of the flux-switching characteristics and larger winding feasibility, the DR-FSDC machine can achieve larger no-load EMF without magnetic saturation, as compared with the DR-DC-MSR machine does.

Fig. 4.5 Calculated meshes
of FEM model.
a DR-DC-MSR machine.
b DR-FSDC machine

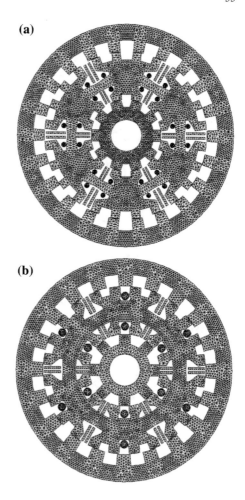

The airgap flux density waveforms of two proposed DR machines under the no-load condition and the DC-field excitation of 700 A-turn, are shown in Fig. 4.8. The peak values of these waveforms are practically equal. In addition, the original fluxes of the outer segment of the stator pole, for both machines, are modulated into four sections in accordance with the number of teeth per stator pole, while of the inner segment of the stator pole are instead modulated into two sections. The results confirm the proposed machines can provide the flux-modulation effect to improve the torque densities.

The DC flux-linkage waveforms at the DSDC mode with the 700 A-turn DC-field current and the self-inductance waveforms at the MSR mode without the DC-field current of the proposed DR machines are shown in Figs. 4.9 and 4.10, respectively. The DC flux-linkages of the DR-DC-MSR machine offers unipolar pattern in both windings as shown in in Fig. 4.9a, while the DC flux-linkages of the DR-FSDC machine offers bipolar pattern as shown in Fig. 4.9b. The results verify

Fig. 4.6 No-load magnetic
field distributions.
a DR-DC-MSR machine.
b DR-FSDC machine

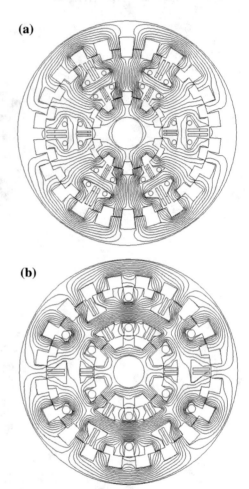

the DR-FSDC machine can provide the flux-switching characteristics. In the
meantime, because the DR-FSDC machine can allocate the larger number of
armature windings as compared with the DR-DR-MSR machine, its
self-inductances are relatively larger accordingly. To retain the same torque value
between the DSDC mode and the MSR mode, the armature current at the MSR
mode can be calculated by Eq. (4.7). The values of K_{DSDC} and K_{MSR} can be
deduced from Figs. 4.9 and 4.10, respectively, while all results are summarized in
Table 4.2.

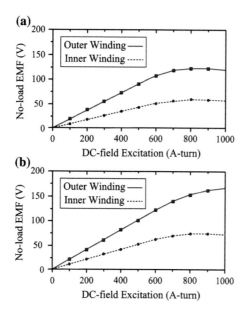

Fig. 4.7 No-load EMF versus DC-field excitation characteristics. **a** DR-DC-MSR machine. **b** DR-FSDC machine

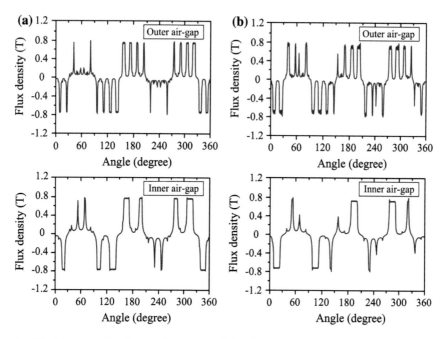

Fig. 4.8 Airgap flux density waveforms. **a** DR-DC-MSR machine. **b** DR-FSDC machine

Fig. 4.9 DC flux-linkage
waveforms at DSDC mode.
a DR-DC-MSR machine.
b DR-FSDC machine

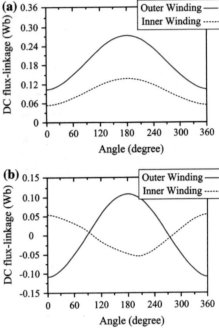

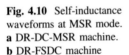 **Fig. 4.10** Self-inductance
waveforms at MSR mode.
a DR-DC-MSR machine.
b DR-FSDC machine

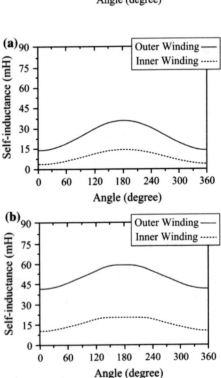

Table 4.2 Key parameters between two operating modes of DR machines

Item	DR-DC-MSR		DR-FSDC	
	Outer	Inner	Outer	Inner
Rated armature current at DSDC mode (A)	5	5	5	5
Slope of DC flux-linkage, K_{DSDC}	0.143/120	0.073/120	0.179/120	0.076/120
Slope of self-inductance, K_{MSR}	0.021/120	0.008/120	0.018/120	0.009/120
Rated armature current at MSR mode (A)	11.7	13.5	14.1	13.0

4.5   Machine Performance Analysis

With the employment of the FEM, the performances of the proposed DR machines can be calculated and compared. First, the no-load EMF waveforms of the DR-DC-MSR machine and DR-FSDC machine are given and as shown in Fig. 4.11. To be specific, the DR-DC-MSR machine can result approximately 121.3 V at the outer winding and 53.4 V at the inner winding. In the meantime, the DR-FSDC machine can result 142.5 V at the outer winding and 71.8 V at the inner winding. The no-load EMF waveforms of two machines consist of the balanced three-phase characteristics. The outer and inner windings of the DR-DC-MSR machine both share the same polarity, while the windings of the DR-FSDC

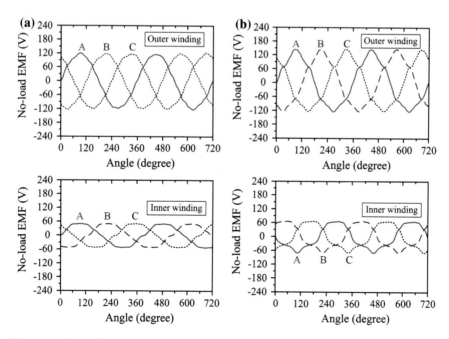

Fig. 4.11 No-load EMF waveforms. **a** DR-DC-MSR machine. **b** DR-FSDC machine

machine instead show the opposite polarities. Since the no-load EMF waveforms of the DR-FSDC machine consists of asymmetrical patterns, larger torque pulsation can be expected. It should be noted that the torque pulsation can be minimized by the injected-harmonic-current technique [20], yet this is out of the scope of this chapter.

Second, the steady torque performances of the DR-DC-MSR machine and the DR-FSDC machine are shown in Figs. 4.12 and 4.13, respectively. The average steady torques of the outer rotor, the inner rotor and the resultant torques of the

Fig. 4.12 Torque waveforms of DR-DC-MSR machine. **a** DSDC mode. **b** MSR mode

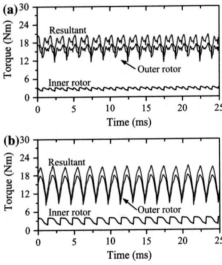

Fig. 4.13 Torque waveforms of DR-FSDC machine. **a** DSDC mode. **b** MSR mode

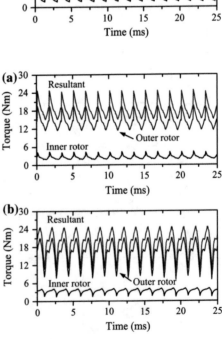

DR-DC-MSR machine at the DSDC mode are 14.9, 2.95 and 17.9 N m, respectively, while at the MSR mode, the values are 14.5, 2.72 and 17.2 N m, respectively. On the other hand, the average steady torques of the outer rotor, the inner rotor and the resultant torques of the DR-FSDC machine at the DSDC mode are 15.8, 3.17 and 19.0 N m, respectively, while at the MSR mode, the values are 15.6, 2.81 and 18.4 N m, respectively. The results confirm that the proposed machines can generate the practically same torque values between two modes. As suggested, the armature currents of two windings of the DR-FSDC machine have to be regulated separately, so that both of its rotors can be operated in the same direction. Furthermore, the resultant torque densities of the DR-DC-MSR machine and the DR-FSDC machine can reach 3635 and 3859 N m/m^3, respectively, and these values are comparable to that generated by (3500–5500 N m/m^3) the common PM machines [1, 2]. In addition, the corresponding torques per current density can reach 3.34 and 3.42 MN m^3/A, respectively, and these values are comparable to that generated by (3.8–5.0 MN m^3/A) the common PM machines as well. It should also be noted the torque pulsating values of the DR-DC-MSR machine and the DR-FSDC machine at the DSDC mode can be found to be 36.3 and 57.3%, respectively, while at the MSR mode the values are 72.4 and 89.3%, respectively. As explained, the torque pulsating values at the MSR mode are higher than that produced at the DSDC mode. Hence, the MSR mode should be employed as the fault-tolerant mode when there are faults on the DC-field currents.

Finally, the cogging torque waveforms of two machines at the DSDC mode with 700 A-turn DC-field excitation are calculated and as shown in Fig. 4.14. The peak values of the outer rotor and the inner rotor of the DR-DC-MSR machine are approximately 0.67 and 0.19 N m, respectively, while the values of the DR-FSDC machine are 0.65 and 0.27 N m, respectively. The generated cogging torques, as

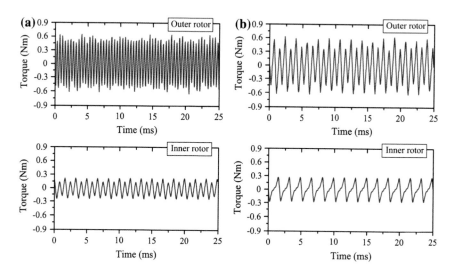

Fig. 4.14 Cogging torque waveforms. **a** DR-DC-MSR machine. **b** DR-FSDC machine

compared with the corresponding rated torques, consist of only 4.5 and 6.4% for the DR-DC-MSR machine and 4.1 and 8.4% for the DR-FSDC machine, and both of these results are within the acceptable range [13].

4.6 Summary

In this chapter, two magnetless DR machines, namely the DR-DC-MSR machine and DR-FSDC machine, are newly proposed and developed, purposely for special direct-drive applications such as driving two separated wheels for EV application or capturing the wind energy with two separated wind blades. The proposed machines can provide two principle of operations, namely the DSDC mode for normal operation and the MSR mode for fault-tolerant operation. Therefore, with the higher fault tolerability, better system reliability can be achieved. With the support of the FEM, the key performances of the two machines are thoroughly analyzed, while the comparisons are summarized in Table 4.3. Among two of them, the DR-FSDC machine provides generally better performances than the DR-DC-MSR counterpart does, in particular regarding the torque density and power density. In the meantime, as suggested, two proposed machines achieve much higher efficiencies at the DSDC mode than that at the MSR mode. It can be confirmed that the proposed DR machines should offer great potentials for EV application and RE generation.

Table 4.3 Double-rotor machine performance comparisons

Item	DR-DC-MSR		DR-FSDC	
	DSDC	MSR	DSDC	MSR
Power (w)	1220	1160	1480	1420
Operating frequency (Hz)	200	200	200	200
Rated speed of outer rotor (rpm)	545	545	600	600
Rated speed of inner rotor (rpm)	1200	1200	1500	1500
No-load EMF of outer rotor (V)	121.3	N/A	142.5	N/A
No-load EMF of inner rotor (V)	53.4	N/A	71.8	N/A
Rated torque of outer rotor (N m)	14.9	14.5	15.8	15.6
Rated torque of inner rotor (N m)	2.95	2.72	3.17	2.81
Resultant rated torque (N m)	17.9	17.2	19.0	18.4
Torque density (N m/m^3)	3635	3493	3859	3737
Torque per current density MN m^3/A	3.34	1.28	3.42	1.22
Rated efficiency	86.4%	68.2%	85.3%	65.8%
Torque ripple of outer rotor	35.1%	70.2%	58.2%	87.2%
Torque ripple of inner rotor	45.6%	74.6%	52.6%	91.6%
Resultant torque ripple	36.3%	72.4%	57.3%	89.3%
Cogging torque of outer rotor	4.5%	N/A	4.1%	N/A
Cogging torque of inner rotor	6.4%	N/A	8.4%	N/A

References

1. M. Cheng, W. Hua, J. Zhang, W. Zhao, Overview of stator-permanent magnet brushless machines. IEEE Trans. Ind. Electron. **58**(11), 5087–5101 (2011)
2. K.T. Chau, C.C. Chan, C. Liu, Overview of permanent-magnet brushless drives for electric and hybrid electric vehicles. IEEE Trans. Ind. Electron. **55**(6), 2246–2257 (2008)
3. Z.Q. Zhu, Switched flux permanent magnet machines – Innovation continues, in *International Conference on Electrical Machines and Systems,* Beijing, China, pp. 1–10, August 2011
4. A. Kawamura, N. Hoshi, T.W. Kim, T. Yokoyama, T. Kume, Analysis of anti-directional-twin-rotary motor drive characteristics for electric vehicles. IEEE Trans. Ind. Electron. **44**(1), 64–70 (1997)
5. M.J. Hoeijmakers, J.A. Ferreria, The electric variable transmission. IEEE Trans. Ind. Appl. **42**(4), 1092–1100 (2006)
6. C. Liu, K.T. Chau, Electromagnetic design and analysis of double-rotor flux-modulated permanent-magnet machines. Prog. Electromagnet. Res. **131**, 81–97 (2012)
7. J.D. Booker, P.H. Mellor, R. Wrobel, D. Drury, A compact, high efficiency contra-rotating generator suitable for wind turbines in the urban environment. Renew. Energy **35**(9), 2027–2033 (2010)
8. C.H.T. Lee, K.T. Chau, C. Liu, D. Wu, S. Gao, Quantitative comparison and analysis of magnetless machines with reluctance topologies. IEEE Trans. Magn. **49**(7), 3969–3972 (2013)
9. K.T. Chau, M. Cheng, C.C. Chan, Nonlinear magnetic circuit analysis for a novel stator doubly fed doubly salient machine. IEEE Trans. Magn. **38**(5), 2382–2384 (2002)
10. Y. Tang, J.J.H. Paulides, T.E. Motoasca, E.A. Lomonova, Flux-switching machine with DC excitation. IEEE Trans. Magn. **48**(11), 3583–3586 (2012)
11. J.T. Chen, Z.Q. Zhu, D. Howe, Stator and rotor pole combination for multi-tooth flux-switching permanent-magnet brushless AC machines. IEEE Trans. Magn. **44**(12), 4659–4667 (2008)
12. R. Cao, C. Mi, M. Cheng, Quantitative comparison of flux-switching permanent-magnet motors with interior permanent magnet motor for EV, HEV, and PEV. IEEE Trans. Magn. **48**(8), 2374–2384 (2012)
13. M. Cheng, K.T. Chau, C.C. Chan, Static characteristics of a new doubly salient permanent magnet motor. IEEE Trans. Energy Convers. **16**(1), 20–25 (2001)
14. C. Liu, K.T. Chau, J.Z. Jiang, A permanent-magnet hybrid brushless integrated-starter-generator for hybrid electric vehicles. IEEE Trans. Ind. Electron. **57**(12), 4055–4064 (2010)
15. W. Zhao, M. Cheng, W. Hua, H. Jia, R. Cao, Back-EMF harmonic analysis and fault-tolerant control of flux-switching permanent-magnet machine with redundancy. IEEE Trans. Ind. Electron. **58**(5), 1926–1935 (2011)
16. W. Li, K.T. Chau, Analytical field for linear tubular magnetic gears using equivalent anisotropic magnetic permeability. Prog. Electromagnet. Res. **127**, 155–171 (2012)
17. Y. Wang, K.T. Chau, C.C. Chan, J.Z. Zhang, Transient analysis of a new outer-rotor permanent-magnet brushless DC drive using circuit-field-torque time-stepping finite element method. IEEE Trans. Magn. **38**(2), 1297–1300 (2012)
18. S.J. Salon, *Finite Element Analysis of Electrical Machines* (Kluwer Academic Publishers, Boston, USA, 1995)
19. S. Niu, S.L. Ho, W.N. Fu, J. Zhu, Eddy current reduction in high-speed machines and eddy current loss analysis with multislice time-stepping finite-element method. IEEE Trans. Magn. **48**(2), 1007–1010 (2012)
20. W. Zhao, M. Cheng, R. Cao, J. Li, Experimental comparison of remedial single-channel operations for redundant flux-switching permanent-magnet motor drive. Prog. Electromagnet. Res. **123**, 189–204 (2012)

Chapter 5
Development of Singly Fed Mechanical-Offset Machine for Torque Ripple Minimization

5.1 Introduction

Energy crisis and environmental pollution have gained more attentions in recent years, so that the developments of renewable energy systems, such as electric vehicle (EV) or wind power generation, have become very popular [1–3]. As the key part of these system, the electric machines have to provide several features, namely high efficiency, high power density, high controllability, wide-speed range, maintenance-free operation and fault-tolerant capability [4–6]. Doubly salient permanent-magnet (DSPM) machines, which can fulfill most of the mentioned criteria, have drawn many attentions in the past few decades [7, 8]. However, even the PM machines can offer a great potential for many industrial applications, they suffer from the disadvantages of high PM material costs and ineffective PM flux regulations [9, 10]. On the other hand, the cost-effective and flux-controllable magnetless doubly salient DC-field (DSDC) machines can relieve the inherited demerits of PM machines such that these machine types have become more popular recently [11, 12].

Without installation of any high-energy-density PM material, the magnetless machines definitely suffer from the demerit of relatively lower torque densities [13–15]. Therefore, the researches on torque density improvement have become one of the hottest issues for these machine types. Meanwhile, as another major component to determine the machine performance, the development of torque ripple minimization has also attracted many attentions [16, 17]. The skewed-rotor structure, which can minimize the torque pulsation problem, has been confirmed to be very effective [18]. However, the conventional skewed-rotor machine installs with only one set of armature windings. Consequently, its excitation partially misaligns with the skewed-rotor positions. To improve the situation, the concept of mechanical-offset (MO) arrangement, which can purposely align the excitations with the skewed-rotor positions, has been developed [19]. However, the asymmetrical torque components produced by concentric machine are undesirable for

© Springer Nature Singapore Pte Ltd. 2018
C. H. T. Lee, *Design, Analysis and Application of Magnetless Doubly Salient Machines*, Springer Theses,
https://doi.org/10.1007/978-981-10-7077-8_5

torque ripple compensation. In addition, the conventional MO machine needs to be operated with the doubly-fed (DF) structure, i.e., with two independent sets of inverters. By taking the cost effectiveness and control simplicity into considerations, unless for fault tolerant topology [20], the DF structure is not favorable in general applications.

This chapter aims to implement the electrical-offset (EO) concept into the MO machine. Consequently, a new singly fed mechanical-offset (SF-MO) machine for torque ripple minimization is formed. As derived from the cascade structure [21], the proposed machine can generate two identical torque components from its two cascaded segments. Hence, a very smooth resultant torque can be potentially generated. Furthermore, the implementation of EO concept can purposely resume the mismatched conducting phases to their original positions. As a result, the proposed MO machine can be operated with the SF configuration. The key machine performances will be analyzed thoroughly based on the finite element method (FEM), while the experimental prototype is developed for verification.

5.2 Proposed Singly Fed Mechanical-Offset Machine

5.2.1 Mechanical-Offset Machine

Figure 5.1 shows the conventional concentric MO machine where its outer- and inner-rotors are purposely mismatched with a conjugated angle θ_m. With this special settlement, the local maxima and local minima of two torque components from two torque producing segments can be compensated with each other. Consequently, relatively smoother resultant torque can be accomplished [19]. Yet,

Fig. 5.1 Conventional MO machine with concentric structure

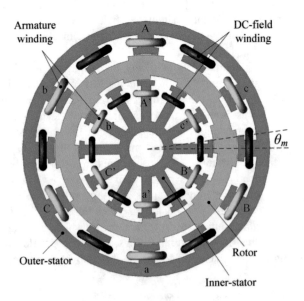

two machine segments based on the concentric topology result with different structures, namely different circumferences and dimensions. As a result, the generated torque components from two segments are different in nature. This particular feature is unfavorable for torque ripple minimization, while improvement can be made if two identical torques can be instead generated.

To improve the situation that exists in the concentric MO machine, the cascade MO machine is proposed in Fig. 5.2. Unlike the concentric machine that decouples two machine segments along its radial direction, the cascade one instead decouples them along its axial direction. Based on the cascade structure, two machine segments theoretically share the same parameters. Hence, the generated torque components should be identical. As a result, two identical torque components can favorably compensate with each other to generate the smoother resultant torque. Nevertheless, similar as the conventional concentric MO machine, the two machine

Fig. 5.2 Proposed MO machine with cascade structure. **a** Machine. **b** Rotor

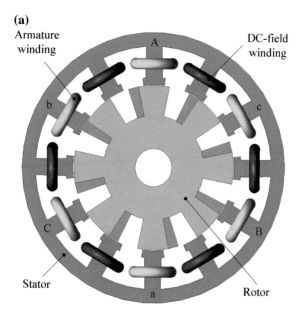

(a)

Armature winding

A

DC-field winding

b

c

C

B

Stator

a

Rotor

(b)

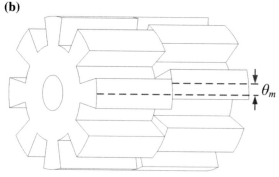

θ_m

segments of the proposed cascade MO machine also need to be controlled separately. Therefore, two independent inverter sets have to be employed and this machine type should be regarded as the DF-MO machine.

The proposed MO machine installs with two types of windings, namely armature winding and DC-field winding. Both of these two windings are wound with the concentrated winding arrangement on alternating stator poles. To reduce the strand induced eddy current and circulating current losses, the litz wire can be employed for armature windings [22]. Nevertheless, the application of the litz wire will increase material costs and decrease winding slot fill factor. Various detailed discussions regarding this matter have been literally available, yet they are out of the scope of this chapter. Hence, the analysis of MO machine with the litz wire will potentially be our future research topic.

5.2.2   Basic Conduction Algorithm

When the DC-field excitation is presence, the conventional DSDC machine can be operated with the bipolar conduction algorithm [6]. The DSDC machine can be purposely designed to provide the sinusoidal-like no-load electromotive force (EMF) waveforms to favor the brushless AC (BLAC) operation. Consequently, torque components with lower torque ripple can be accomplished based on this arrangement. To illustrate the basic conduction algorithm, the DSDC machine is purposely extended to become the cascade form, so-called as the cascade DSDC (C-DSDC) machine. The C-DSDC machine can also be regarded as the non-skewed-rotor machine. To be specific, its two machine segments are aligned with each other and this arrangement is similar to those employed in the double-stator (DS) machine [17].

To produce the positive electromagnetic torque, the sinusoidal armature current I_{BLAC} is employed in accordance to the status of flux-linkage Ψ. Because the tooth pairs of two machine segments of the C-DSDC machine align with each other, its two armature winding sets can be conducted simultaneously with no phase shift. In particular, as shown in Fig. 5.3a, the three-phase topology is selected and its two armature currents can be described as

$$\begin{cases} i_1 = I_{\max} \sin(\theta + \theta_k) \\ i_2 = I_{\max} \sin(\theta + \theta_k) \end{cases} \tag{5.1}$$

where i_1 and i_2 are two armature winding sets, θ_k is initial angle with value of $2\,k\pi/m$, m is number of armature phases and k is any integer. With this conduction scheme, the powers from two stators can be transferred to rotor simultaneously. Two armature windings can be connected in series to reduce the operating complexity. Consequently, this machine type can be regarded as the SF machine.

Fig. 5.3 Theoretical
operating waveforms.
a C-DSDC machine.
b DF-MO-DSDC machine.
c SF-MO-DSDC machine

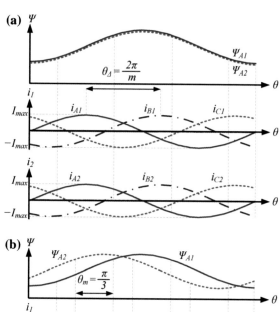

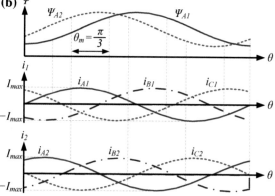

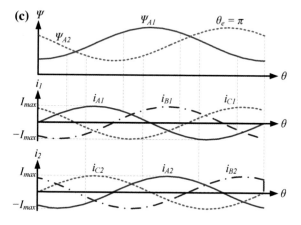

However, the local maxima and local minima of two torque components are integrated with each other in an undesirable way such that the torque ripple value is very large.

5.2.3 Existing Mechanical-Offset Conduction Algorithm

Based on the MO design, two tooth pairs of the DF-MO-DSDC machine are purposely offset with a conjugated angle [19]. Consequently, the conduction angles of two torque producing segments are mismatched with $\theta_m = \pi/m$, as shown in Fig. 5.3b. Therefore, to drive the machine properly, two armature winding sets of the three-phase DF-MO-DSDC machine need to be controlled independently as

$$\begin{cases} i_1 = I_{\max} \sin(\theta + \theta_k) \\ i_2 = I_{\max} \sin(\theta + \theta_k + \theta_m) \end{cases} \tag{5.2}$$

With this mismatched settlement, the local maxima and local minima of two torque components can be favorably superimposed with each other. As a result, the torque ripple minimization can be accomplished. It should be noted its average torque value should be retained to the same level as compared with that generated by the C-DSDC machine.

Unlike the basic C-DSDC machine that can connect two armature windings in series, the DF-MO-DSDC machine needs to conduct its two armature windings individually. Consequently, the conducting currents can response to the corresponding Ψ from two machine segments accordingly. The DF-MO-DSDC machine can therefore be regarded as the skewed-rotor machine with the DF excitation. As a result, this MO structure has no choice but to increase the number of conducting phases, so as control complexity and cost of power electronics.

5.2.4 Proposed Electrical-Offset Algorithm

The conventional MO arrangement suffers from the undesirable control complexity, yet the situation can be improved by engagement of the proposed EO algorithm. Because of the repetitive feature of trigonometry, the so-called EO angle θ_e can be purposely integrated into the MO structure. Consequently, the shifted phases can be resumed to its original positions. Based on this proposed settlement, the armature winding currents can be described as

$$\begin{cases} i_1 = I_{\max} \sin(\theta + \theta_k) \\ i_2 = I_{\max} \sin(\theta + \theta_k + \theta_m + \theta_e) \end{cases} \tag{5.3}$$

To resume the shifted phases to the original positions, the corresponding angles need to fulfill the criterion as follows

Table 5.1 Electrical-offset design combinations

m	θ_m (°)	$\theta_m + \theta_e$ (°)	θ_Δ (°)	m_s
3	60	240	120	2
5	36	216	72	3
7	25.7	205.7	51.4	4
9	20	200	40	5

$$\theta_k = \theta_m + \theta_e \tag{5.4}$$

By moving θ_e as major term in Eq. (5.4), the relationship can be further deduced as

$$\theta_e = \frac{(2k-1)\pi}{m} \tag{5.5}$$

For $m = 2k$, i.e., the machine with even number of armature phases, the relationship ends up with infinitely many solutions and it is not possible for realization. On the other hand, for $m = 2k - 1$, i.e., the machine with odd number of armature phases, the relationship results with a particular solution as $\theta_e = \pi$. With the implementation of the proposed EO algorithm, the shifted phases can be resumed to its original positions and the number of phase shifted m_s is governed by

$$m_s = \frac{\theta_m + \theta_e}{\theta_\Delta} = \frac{m+1}{2} \tag{5.6}$$

where θ_Δ is angle difference between armature phases. Based on Eqs. (5.5) and (5.6), the design combinations of the proposed EO algorithm can be found in Table 5.1.

Based on the EO concept, the shifted phases can be resumed to match with the original conduction angles, as shown in Fig. 5.3c. Therefore, the torque components from two segments can be superimposed with each other to generate a smoother torque. In the meantime, the number of conducting phases and the requirement of power electronics can be minimized. It should be noted that the proposed SF-MO-DSDC machine can be regarded as the skewed-rotor machine with the SF excitation.

5.2.5 Proposed Machine Structure

The machine structure of the proposed SF-MO-DSDC magnetless machine, which consists of 12 stator poles and 8 rotor poles, is shown in Fig. 5.2. Because the proposed SF-MO-DSDC machine is extended from the DSDC machines, its design equations can be deduced from the conventional DSDC machines as [6]

$$\begin{cases} N_s = 2mj \\ N_r = N_s \pm 2j \end{cases} \tag{5.7}$$

where N_s is number of stator poles, N_r is number of rotor poles and j is any integer. To simplify the control complexity and to minimize the cost of power electronic devices, the least number of armature phases, i.e., three-phase structure, is selected. In addition, to improve the torque density and to minimize the torque ripple value, repetitive tooth structure should be chosen. By taking these criteria into considerations, the designs of $j = 2$, $m = 3$, $N_s = 12$ and $N_r = 8$ are selected for the proposed SF-MO-DSDC machine.

In order to accomplish the proposed EO concept, the SF-MO-DSDC machine needs to engage with two mismatched angles, namely the mechanical θ_m and the electrical θ_e. To be specific, the θ_m can be actualized by mismatching two rotor segments with the mechanical displacement of 7.5°, i.e., electrical angle of $\pi/3$, as shown in Fig. 5.2b. Meanwhile, the θ_e can be actualized by connecting two stator segments with opposite excitation polarities. Consequently, two segments can response with an electrical displacement with π. This opposite polarities settlement can be easily realized by connecting the DC-field winding sets with a reverse direction.

5.2.6   Proposed System Arrangement

The system arrangements of the DF-MO-DSDC machine and the proposed SF-MO-DSDC machine are shown in Fig. 5.4. The systems consist of three key parts, namely (i) the armature inverters, (ii) the H-bridge converters and (iii) the machines. The armature inverters can produce the appropriate BLAC currents to drive the machines, while the H-bridge converters can regulate magnitudes and directions of the DC-field excitations.

As aforementioned, the conventional MO machine needs to decouple its two armature winding sets to accomplish torque ripple minimization. Therefore, to drive the DF-MO-DSDC machine properly, two individual inverter sets are needed. In the meantime, with the implementation of the EO concept, the shifted conducting phases of the MO machine can be resumed to its original conduction positions. Unlike the DF-MO-DSDC machine that connects the DC-field winding sets with same polarity, the SF-MO-DSDC machine purposely connects the DC-field winding sets with opposite polarities. As a result, as described in Eq. (5.6), two armature winding sets should be connected as the following arrangements: A-phase connects with C'-phase in series, B-phase with A'-phase and C-phase with B'-phase. With these winding settlements, the proposed SF-MO-DSDC machine can be operated with only one inverter set.

(a)

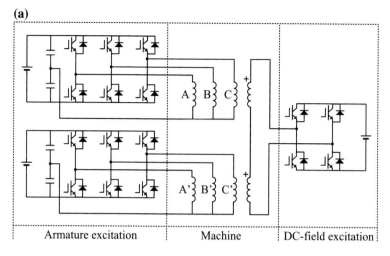

Armature excitation   Machine   DC-field excitation

(b)

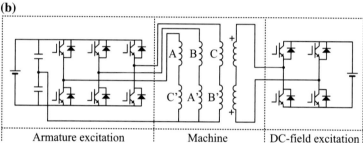

Armature excitation   Machine   DC-field excitation

Fig. 5.4 System arrangement. **a** DF-MO-DSDC. **b** SF-MO-DSDC

5.3 Machine Performance Analysis

5.3.1 Electromagnetic Field Analysis

Regarding the electric machine analysis, the FEM-based electromagnetic field analysis has been well agreed as the most convenience and accurate tool [6]. In this chapter, a commercial FEM software package, JMAG-Designer is adopted for performance analysis. Hence, the major machine dimensions and the key parameters can be optimized upon iterative approach.

With the existence of DC-field excitation of 10 A/mm^2, the flux-linkage waveforms of the proposed machine at base speed of 3500 rpm are shown in Fig. 5.5. Because the two cascaded segments of the proposed machine consist of identical structure, to improve the readability, only one set of flux-linkage waveforms is shown. It can be shown that the proposed machine can produce the well-balanced flux-linkages among three-phase characteristics with no significant distortion.

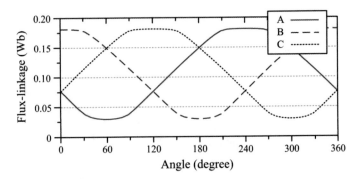

Fig. 5.5 Flux-linkage waveforms of the proposed machine

5.3.2 Pole-Arc Ratio Analysis

The no-load EMF has been regarded as one of the major criteria to determine the machine performances and it should be analyzed thoroughly. To be specific, the so-called the pole-arc ratio p, which is defined as the ratio of rotor pole-arc p_r to stator pole-arc p_s, i.e., $p = p_r/p_s$ is carefully studied for dimension optimization. To optimize between magnetic saturation and winding slot area, at the beginning stage, p_s is set with an initial value. First, β_r is selected to be the same value as p_s, i.e., $p = 1$, as shown in Fig. 5.6a. Next, p_r is adjusted in a way that the optimal pole-arc ratio $p\_opt = p_{r\_opt}/p_s$ can be accomplished, as shown in Fig. 5.6b. Based on the operating conditions of DC-field excitation of 10 A/mm^2 and speed of 3500 rpm, the variations of no-load EMF waveforms in according to various p are shown in Fig. 5.7. As previously suggested, the proposed machine should be designed to produce no-load EMF waveform with more sinusoidal-like characteristic. Consequently, relatively smoother torque components can be generated based on the BLAC conduction scheme. Based on this argument, the pole-arc ratio should be selected between $p = 1.4$ and 1.5.

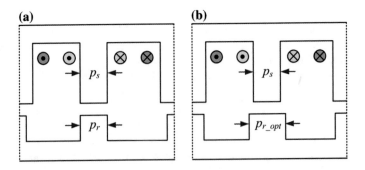

Fig. 5.6 Pole-arc ratio optimization. **a** Initial case. **b** Optimal case

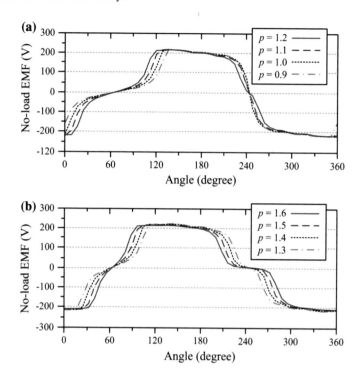

Fig. 5.7 No-load EMF waveforms under various pole-arc ratios. **a** p = 0.9–1.2. **b** p = 1.3–1.6

Apart from the no-load EMF characteristics, the cogging torque is another major component that should be studied carefully. Based on the mentioned operating conditions, the cogging torque waveforms under different values of p are shown in Fig. 5.8. When $p = 1.4$ and 1.5, the peak values of cogging torque become about 8.4 and 7.6 N m, respectively. In general, when lower cogging torque is accomplished, smaller torque ripple and better machine performances can then be generated. Hence, to provide the most desirable no-load EMF waveform with the lowest cogging torque value, the optimal pole-arc ratio is chosen as $p\_{opt} = 1.5$.

To offer a more comprehensive analysis on the selected pole-arc ratio, a sensitivity analysis of no-load EMF and cogging torque based on various DC-field excitations I_{dc} is conducted in Fig. 5.9. As illustrated, the DC-field excitation can affect the magnitudes of these quantities while it has almost no effect on their patterns. Hence, it can be confirmed the optimal pole-arc ratio retains the same even with various DC-field excitations.

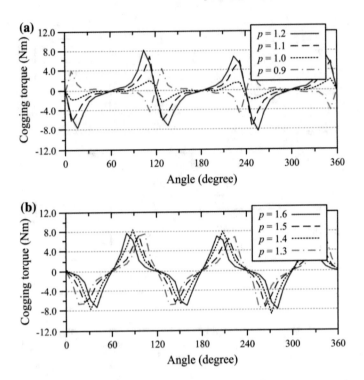

Fig. 5.8 Cogging torques under various pole-arc ratios. **a** p = 0.9–1.2. **b** p = 1.3–1.6

5.3.3 No-Load EMF Analysis

In order to offer a comprehensive evaluation, the C-DSDC machine and the DF-MO-DSDC machine are purposely included for comparisons. To provide a fair environment for comparison, all the major machines dimensions, namely outside diameter, inside diameter, stack length, airgap length and winding slot fill factor are set equal. The corresponding key design data of all machines are listed in Table 5.2.

With the support of FEM, the no-load EMF waveforms of the C-DSDC machine, the DF-MO-DSDC machine and the SF-MO-DSDC machine at base speed of 3500 rpm are shown in Fig. 5.10. Because the two tooth-pairs of the cascaded segments align with each other, two sets of no-load EMF waveforms of the C-DSDC machine superimpose perfectly with each other, as shown in Fig. 5.10a. In the meantime, the DF-MO-DSDC machine purposely displaces its two cascaded segments with a conjugated angle θ_m. Consequently, its two sets of no-load EMF waveforms instead mismatch with an offset angle of $\pi/3$, as shown in Fig. 5.10b.

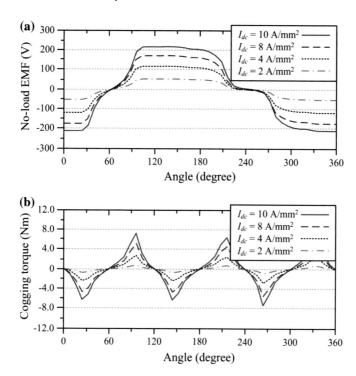

Fig. 5.9 Sensitivity analysis based on various DC-field excitations. **a** No-load EMFs. **b** Cogging torques

Table 5.2 Key design data of the proposed machines

Items	C-DSDC	DF-MO-DSDC	SF-MO-DSDC
Power (kW)	56	56	56
Base speed (rpm)	3500	3500	3500
Stator outside diameter (mm)	312.0	312.0	312.0
Stator inside diameter (mm)	192.0	192.0	192.0
Rotor outside diameter (mm)	190.0	190.0	190.0
Rotor inside diameter (mm)	44.0	44.0	44.0
No. of stator poles	12	12	12
No. of rotor poles	8	8	8
Stator pole arc (°)	15.0	15.0	15.0
Rotor pole arc (°)	22.5	22.5	22.5
Airgap length (mm)	1.0	1.0	1.0
Stack length (mm)	120 * 2	120 * 2	120 * 2
No. of armature turns	14	14	14
Mechanical-offset θ_m (°)	N/A	7.5	7.5
Electrical-offset θ_ε (°)	N/A	N/A	180

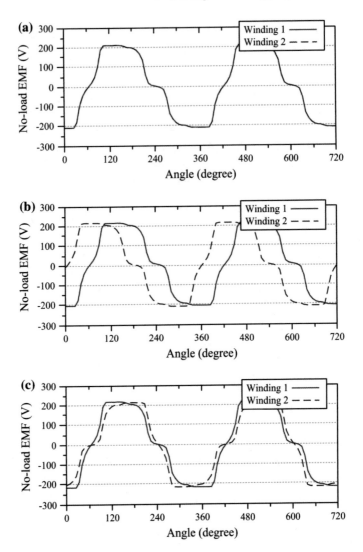

Fig. 5.10 No-load EMF waveforms of two armature winding sets. **a** C-DSDC machine. **b** DF-MO-DSDC machine. **c** SF-MO-DSDC machine

The proposed SF-MO-DSDC machine is derived from the DF-MO-DSDC machine while θ_e is implemented to accomplish the EO arrangement. Then, its two sets of no-load EMF waveforms can be resumed to the original positions, as shown in Fig. 5.10c. Because of the superimposition effect of θ_m and θ_e, the conducting phases of the SF-MO-DSDC machine are shifted with relationship of $m_s = 2$, i.e., A-phase aligns with C′-phase, B-phase with A′-phase and C-phase with B′-phase.

5.3.4 Torque Performances

Based on the mentioned operating conditions, the torque performances of the proposed machine and its counterparts under the BLAC operation are shown in Fig. 5.11. It can be shown that the average steady torques of the C-DSDC machine, the DF-MO-DSDC machine and the SF-MO-DSDC machine are around 155.1, 154.6 and 154.3 N m, respectively. Obviously, the results verify both of the DF-MO-DSDC machine and the SF-MO-DSDC machine can superimpose its two torque components perfectly to generate same torque level, as compared with the basic C-DSDC machine. Moreover, the peak values of all cogging torques are about 15.2 N m and they are only 9.8, 9.8 and 9.9% of their average torques. Therefore, all cogging torque values are perceived as very acceptable, as compared with the well-developed PM counterparts [4, 6].

To provide a more comprehensive study of the torque performances, torque ripple values are also carefully analyzed. In particular, the torque ripples of the C-DSDC machine, the DF-MO-DSDC machine and the SF-MO-DSDC machine are about 72.5, 7.2 and 7.2%, respectively. Undoubtedly, with no torque ripple compensation, the C-DSDC machine suffers from the largest torque pulsation. In the meantime, based on the proposed designs, the local maxima and local minima of torque components from both DF-MO-DSDC machine and SF-MO-DSDC machine are seamlessly integrated with each other. Consequently, the torque ripple values from these two machines can be minimized. Even though there are some machine types that result very low torque ripple values [15], the proposed concept offers an addition direction for exploration.

With the traditional structure, two armature winding sets of the C-DSDC machine can integrate with each other and it can be regarded as the three-phase SF machine. On the other hand, based on the MO design, the DF-MO-DSDC machine needs to displace its two armature winding sets with a conjugated angle θ_m. Consequently, this machine type should be regarded as a six-phase machine, or known as the three-phase DF machine. Meanwhile, the SF-MO-DSDC machine can employ the EO concept to resume the conducting phases to its original positions. As a result, the proposed SF-MO-DSDC machine can be regarded as the three-phase SF machine and it can enjoy the definite merits of simple power electronics structure. The torque performances of the proposed machines are compared and categorized in Table 5.3.

5.4 Experimental Verifications

To testify the proposed concept and simulated results, the experimental setup of the proposed machine is developed and as shown in Fig. 5.12. To achieve sensible and practical experiments in the laboratory, the power level of the prototype is

Fig. 5.11 Steady torque waveforms. **a** C-DSDC machine. **b** DF-MO-DSDC machine. **c** SF-MO-DSDC machine

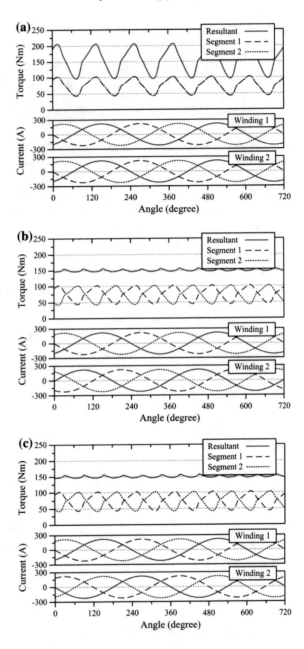

purposely scaled down. The established machine is designed with the outside diameter of 156 mm, stack length of 60 mm and airgap length of 1 mm.

The winding sets of the established machine are purposely decoupled between two cascaded segments, such that it can simultaneously act as two machine

Table 5.3 Torque performances of the proposed machines

Items	C-DSDC	DF-MO-DSDC	SF-MO-DSDC
No. of phases	3	6	3
Average torque (N m)	155.1	154.6	154.3
Cogging torque (N m)	15.2	15.2	15.2
% cogging torque	9.8%	9.8%	9.9%
Torque ripple	72.5%	7.2%	7.2%

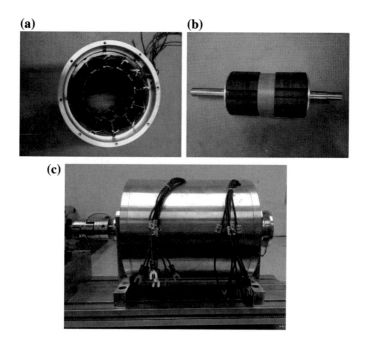

Fig. 5.12 Prototype. **a** Stator. **b** Rotor. **c** Assembled machine

scenarios, namely the DF-MO-DSDC machine scenario and the SF-MO-DSDC machine scenario. To be specific, the former scenario can be actualized if the DC-field winding sets are connected with same polarity. On the other hand, the latter one can instead be actualized if the DC-field winding sets are connected in opposite polarities. Because the two winding sets are decoupled, the end windings are enlarged so that two rotor segments are separated, as shown in Fig. 5.12b.

The measured no-load EMF waveforms of the proposed machine with the DC-field excitation of 3 A/mm^2 and operating speed of 900 rpm are shown in Fig. 5.13. The results show that the measured waveforms can well comply with the calculated waveforms as shown in Fig. 5.10a. To illustrate the effect of the EO implementation, the no-load EMF waveforms of the corresponding armature phases at the DF-MO-DSDC machine scenario and the SF-MO-DSDC machine scenario

Fig. 5.13 Measured no-load
EMF waveforms of the
proposed machine (20 V/div).
a Winding 1. **b** Winding 2

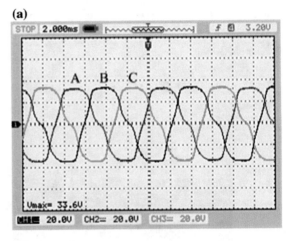

(a)

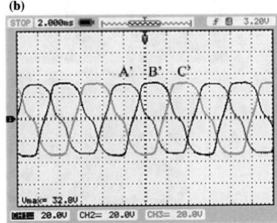

(b)

with the DC-field excitation of 1 A/mm^2 and operating speed of 900 rpm are shown
in Fig. 5.14. It can be shown that the armature phases of the DF-MO-DSDC
machine, namely the A-phase, B-phase, A′-phase and B′-phase are displaced with an
electrical angle of $\pi/3$. Hence, the DF-MO-DSDC machine should be regarded as the
six-phase machine. In the meantime, the armature phases of the SF-MO-DSDC
machine, namely the C′-phase and A′-phase are resumed to align with the corresponding A-phase and B-phase, respectively. Therefore, the SF-MO-DSDC machine
should be regarded as the three-phase SF machine. These measured results well
agree with the calculated waveforms as shown in Fig. 5.10b, c.

Without a dynamic torque transducer, the electromagnetic torque of the prototype cannot be measured directly. Meanwhile, other measured values of the
established prototype can well comply with the calculated waveforms based on the
finite element analysis. To testify the reliability of FEM analysis, the calculated
no-load EMF values with the DC-field excitation of 10 A/mm^2 at various speeds

Fig. 5.14 Measured no-load
EMF waveforms (10 V/div).
a DF-MO-DSDC scenario.
b SF-MO-DSDC scenario

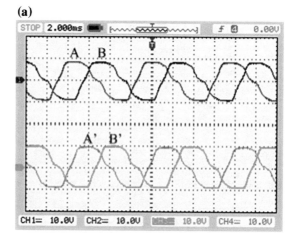

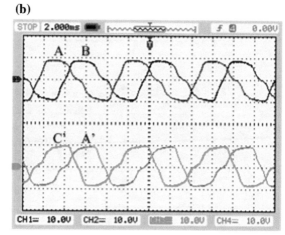

are compared with the experimental results, as shown in Fig. 5.15. As illustrated, the maximum errors between the FEM analysis and experimental results are less than 3%. Therefore, it can be concluded that the calculated torque performances are credible and reliable. Moreover, the difference between two machine scenarios are less than 2%. Consequently, it can be deduced that two machine scenarios are nearly identical.

As one of the major features for renewable energy applications, such as EV applications, the flux-weakening performances of the proposed machine should be analyzed with details. The corresponding variations of the measured back EMF characteristics with respect to the operating speed at no load, without and with flux regulations, are shown in Fig. 5.16. The results verify that the proposed machine at two machine scenarios can both utilize its flux-controllable features to maintain

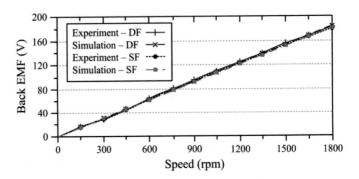

Fig. 5.15 No-load EMF characteristics versus speed at no load

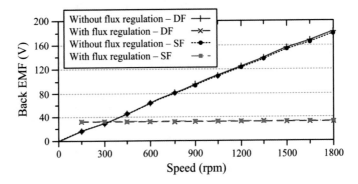

Fig. 5.16 Measured back EMF characteristics versus speed at no load

their back EMFs at a suitable level. Consequently, excellent flux-weakening capability for wide-speed range operation can be accomplished.

Moreover, the back EMF characteristics of the proposed machine with various load conditions are also carefully studied. The corresponding variations of the measured back EMF characteristics with the DC-field excitation of 10 A/mm^2 and the operating speed of 900 rpm with respect to the load currents, without and with flux regulations, are shown in Fig. 5.17. Undoubtedly, the measured back EMFs can be retained at the desirable level. Therefore, the results further testify the proposed machine can provide excellent flux-weakening capability at wide ranges of operating speeds and load currents. This feature is highly favorable for modern EV applications.

Finally, as shown in Fig. 5.18, the efficiencies of the proposed machine with the DC-field excitation of 10 A/mm^2 at two scenarios based on various operating speeds and load currents are also measured. It can be shown that the power varies along the operating speeds and load currents. Meanwhile, the proposed machine can reach 150 W at rated conditions, i.e., at operating speed of 900 rpm and load current of 1.0 A. When the electronic loads of 150 W are employed, the efficiencies

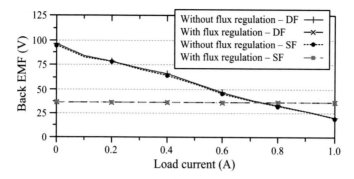

Fig. 5.17 Measured back EMF characteristics versus load current

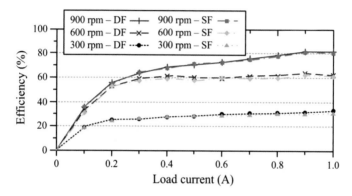

Fig. 5.18 Measured efficiencies under various speeds and load currents

of the DF-MO-DSDC machine scenario and the SF-MO-DSDC machine scenario can reach up to 81 and 80%, respectively. Even the measured efficiencies are not as satisfactory as that produced by PM machines, the overall performances of the proposed machine are quite attractive.

5.5   Summary

A new SF-MO-DSDC machine for renewable applications has been proposed and established in this chapter. By implementing the EO concept, the proposed machine can well superimpose its torque components to suppress the resultant torque pulsation. Different from the traditional DF-MO-DSDC machine that needs two independent inverter sets, the proposed SF-MO-DSDC machine can actualize the

EO angle to resume the shifted conducting phases to its original positions. Therefore, the proposed machine can minimize the costs of power electronics and the control complexity.

References

1. J.P. Barton, D.G. Infield, Energy storage and its use with intermittent renewable energy. IEEE Trans. Energy Convers. **19**(2), 431–438 (2004)
2. A. Emadi, Y.J. Lee, K. Rajashekara, Power electronics and motor drives in electric, hybrid electric, and plug-in hybrid electric vehicles. IEEE Trans. Ind. Electron. **55**(6), 2237–2245 (2008)
3. J.D. Santiago, H. Bernhoff, B. Ekergard, S. Eriksson, S. Ferhatovic, R. Waters, M. Leijon, Electrical motor drivelines in commercial all-electric vehicles: a review. IEEE Trans. Veh. Technol. **61**(2), 475–484 (2012)
4. Z.Q. Zhu, D. Howe, Electrical machines and drives for electric, hybrid, and fuel cell vehicles. Proc. IEEE **95**(4), 746–765 (2007)
5. A. Tenconi, S. Vaschetto, A. Vigliani, Electrical machines for high-speed applications: design considerations and tradeoffs. IEEE Trans. Ind. Electron. **61**(6), 3022–3029 (2014)
6. K.T. Chau, *Electric Vehicle Machines and Drives—Design, Analysis and Application* (Wiley-IEEE Press, New York, 2015)
7. M. Cheng, W. Hua, J. Zhang, W. Zhao, Overview of stator-permanent magnet brushless machines. IEEE Trans. Ind. Electron. **58**(11), 5087–5101 (2011)
8. I.A.A. Afinowi, Z.Q. Zhu, Y. Guan, J.C. Mipo, P. Farah, A novel brushless AC doubly salient stator slot permanent magnet machine. IEEE Trans. Energy Convers. **31**(1), 283–292 (2016)
9. I. Boldea, L.N. Tutelea, L. Parsa, D. Dorrell, Automotive electric propulsion systems with reduced or no permanent magnets: an overview. IEEE Trans. Ind. Electron. **61**(10), 5696–5711 (2014)
10. C.H.T. Lee, K.T. Chau, C. Liu, Design and analysis of a cost-effective magnetless multi-phase flux-reversal DC-field machine for wind power generation. IEEE Trans. Energy Convers. **30**(4), 1565–1573 (2015)
11. D. Dorrell, L. Parsa, I. Boldea, Automotive electric motors, generators, and actuator drive systems with reduced or no permanent magnets and innovative design concepts. IEEE Trans. Ind. Electron. **61**(10), 5693–5695 (2014)
12. C.H.T. Lee, K.T. Chau, C. Liu, C.C. Chan, Overview of magnetless brushless machines. IET Electr. Power Appl. (accepted)
13. Z.Q. Zhu, Z.Z. Wu, D.J. Evans, W.Q. Chu, A wound field switched flux machine with field and armature windings separately wound in double stators. IEEE Trans. Energy Convers. **30**(2), 772–783 (2015)
14. C. Yu, S. Niu, Development of a magnetless flux switching machine for rooftop wind power generation. IEEE Trans. Energy Convers. **30**(4), 1703–1711 (2015)
15. Z.Q. Zhu, Z.Z. Wu, X. Liu, A partitioned stator variable flux reluctance machine. IEEE Trans. Energy Convers. **31**(1), 78–92 (2016)
16. S.H. Han, T.M. Jahns, W.L. Soong, M.K. Guven, M.S. Illindala, Torque ripple reduction in interior permanent magnet synchronous machines using stators with odd number of slots per pole pair. IEEE Trans. Energy Convers. **25**(1), 118–127 (2010)
17. A.H. Isfahani, B. Fahimi, Comparison of mechanical vibration between a double-stator switched reluctance machine and a conventional switched reluctance machine. IEEE Trans. Magn. **50**(2), 7007104 (2014)
18. W. Fei, P.C.K. Luk, J. Shen, Torque analysis of permanent-magnet flux switching machines with rotor step skewing. IEEE Trans. Magn. **48**(10), 2664–2673 (2012)

19. C.H.T. Lee, K.T. Chau, C. Liu, T.W. Ching, F. Li, Mechanical offset for torque ripple reduction for magnetless double-stator doubly salient machine. IEEE Trans. Magn. **50**(11), 8103304 (2014)
20. W. Zhao, M. Cheng, W. Hau, H. Jia, R. Cao, Back-EMF harmonics analysis and fault-tolerant control of flux-switching permanent-magnet machine with redundancy. IEEE Trans. Ind. Electron. **58**(5), 1926–1935 (2011)
21. Y. Wang, Z. Deng, A multi-tooth fault-tolerant flux-switching permanent-magnet machine with twisted-rotor. IEEE Trans. Magn. **48**(10), 2674–2684 (2012)
22. T. Raminosoa, D.A. Torrey, A. El-Refaie, D. Pan, S. Grubic, K. Grace, Robust non-permanent magnet motors for vehicle propulsion, in *Proceedings of IEEE International Electric Machines & Drives Conference* (2015), pp. 496–502

Part II
Design, Analysis and Application of Advanced Magnetless Machines on Wind Power Generations

Chapter 6
Proposed Flux-Reversal DC-Field Machine for Wind Power Generation

6.1 Introduction

Energy crisis and environmental pollution have become the worrying problems since the last few decades, while the renewable wind power generation has confirmed to be one of the most probable solutions to relieve the situations [1, 2]. Generally speaking, the wind power generations can be classified into two major categories, namely the constant-speed constant-frequency (CSCF) wind power generation and the variable-speed constant-frequency (VSCF) wind power generation [3, 4]. Without implementing any power converters, the CSCF type benefits from the absolute predominance of simple system structure and high robustness [5]. Nevertheless, in the CSCF mechanism, the system has to disconnect the response with the wind speed such that the turbine speed has to be confined in a limited range. Not surprisingly, the CSCF system has to bear the high mechanical stress problems and ends up with relatively low efficiency.

To improve the situation from the CSCF system, the VSCF type has been developed accordingly. With the support of the power electronics, the VSCF system can vary its turbine speed in accordance to the wind speed, in order to capture the maximum wind energy for power generation [6, 7]. Undoubtedly, the VSCF type can provide higher overall efficiency as compared with the CSCF one. As the soul of the VSCF system, same as the electric vehicle (EV) applications, the electric machines have to fulfill several criteria, including high efficiency level, high power density, high controllability, wide-speed operating range and brushless operation [8]. The doubly salient permanent-magnet (DSPM) machines, which can achieve most of the mentioned goals, have been actively developed. Particularly, the flux-reversal PM (FRPM) machine, which offers the bipolar flux-linkage and thus higher power density, has attracted more attentions recently [9]. However, similar to the well-developed DSPM machines, the FRPM machine also has to tackle two vital problems, namely the high PM material cost and the uncontrollable PM flux densities [10]. To get round these structural problems, the doubly salient DC-field

© Springer Nature Singapore Pte Ltd. 2018
C. H. T. Lee, *Design, Analysis and Application of Magnetless Doubly Salient Machines*, Springer Theses,
https://doi.org/10.1007/978-981-10-7077-8_6

(DSDC) machine was suggested. By replacing the PM materials with the independent DC-field winding, the DSDC machine can enjoy the benefits of higher cost-effectiveness and better flux controllability. Yet, the development of the FRDC machine is still absent.

This chapter aims to implement the new modulated stator pole structure for the specific winding arrangement, to emulate the PM configuration that exists in the FRPM machines. Hence, a new magnetless four-phase FRDC machine is proposed, purposely for the wind power generation system. The machine performances are analyzed by the finite element method (FEM), while the experimental results are also provided to verify the proposed idea.

6.2 System Architecture

Figure 6.1 shows the wide-speed range wind power generation system configuration. To be specific, it consists of seven key elements: (i) a wind turbine equipped with gearbox to capture the wind power, (ii) the proposed FRDC machine to converse the electromechanical energy, (iii) a full-bridge rectifier to provide the AC-DC conversion, (iv) a battery to store the generated energy, (v) an inverter to provide the AC-DC conversion to the power grid, (vi) a DC-field controller to feedback the DC-field current signal and (vii) a buck converter to regulate the DC voltage for flux regulation.

Based on the Betz theory, the mechanical power captured by the wind turbine can be described as [6]

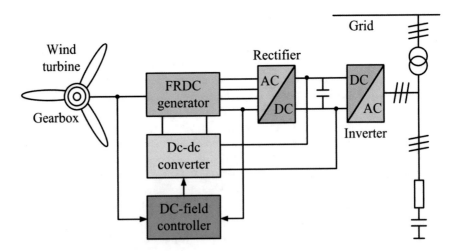

Fig. 6.1 Wind power generation system

$$P_{mech} = \frac{1}{2} C_p \rho v_\omega^3 A \qquad (6.1)$$

where C_p is the coefficient of wind power conversion factor with a typical value of 0.4 or below, ρ is the air density, v_ω is the wind velocity and A is the swept area of the wind-turbine rotor. The value of C_p is a function based on the ratio of the blade tip speed to the wind speed β and defined as

$$\beta = \frac{\omega R}{v_\omega} \qquad (6.2)$$

where R is the radius of the blades and ω is the rotational speed of the wind-turbine shaft. For a specific value β_{max}, the power conversion factor C_p can reach to a single maxima in order to capture the maximum mechanical power from the wind speed. Undoubtedly, the turbine speed should vary along with the wind speed to maximize its total power generation. With turbine runs at the β_{max}, the maximum generated power can then be described as

$$P_{mech} = \frac{1}{2} \left(C_{pmax} \pi R^2 \rho \right) v_\omega^3 \qquad (6.3)$$

According to Eqs. (6.2) and (6.3), the maximum generated power can be further described as

$$P_{mech} = \frac{1}{2} \left(\frac{C_{pmax} \pi R^5 \rho}{\beta_{max}^3} \right) \omega^3 \qquad (6.4)$$

It should be noted that the turbine speed of the wind generator is typical around 1000 rpm at the normal wind situation.

6.3 Proposed Flux-Reversal DC-Field Machine

6.3.1 Proposed Machine Structure

The topologies of the traditional FRPM machine and the proposed four-phase FRDC machine are shown in Fig. 6.2. Because the proposed FRDC machine is developed based on the typical FRPM machines, its design equations such as the pole arrangements can be extended from that of the traditional FRPM machines and as shown below

Fig. 6.2 Machine structures.
a FRPM type. **b** Proposed
FRDC type

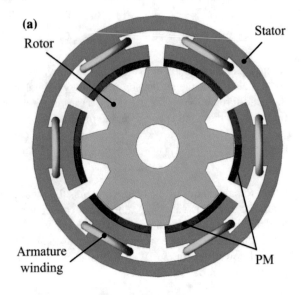

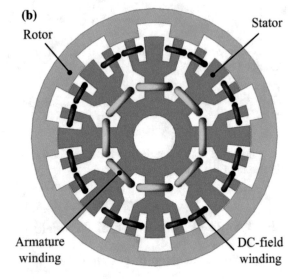

$$\begin{cases} N_{sp} = 2mk \\ N_{se} = N_{sp}p_s \\ N_r = N_{se} + 2k \end{cases} \tag{6.5}$$

where N_{sp} is the number of actual stator poles, p_s the flux pole-pair on the stator tooth, N_{se} the apparent stator poles, N_r the rotor poles, m the armature phases and k any integer. The fundamental design combinations of the proposed FRDC

Table 6.1 Fundamental design combinations of the FRDC machine

m	k	N_{sp}	p_s	N_{se}	N_r
3	1	6	1	6	8
3	2	12	1	12	16
3	1	6	2	12	14
4	1	8	1	8	10
4	2	16	1	16	20

machine are categorized in Table 6.1. Based on the considerations of the stability and cost-effectiveness, the four-phase topology is preferred. Meanwhile, to ease the manufacturing complexity, the minimum number of flux pole-pairs should be chosen. By taking these concerns into consideration, the following combination is selected: $m = 4$, $k = 1$, $N_{sp} = 8$, $p_s = 1$, $N_{se} = 8$ and $N_r = 10$, as the proposed structure for the FRDC machine.

The traditional FRPM machine equips the PMs of alternating polarities on each of its stator pole, and hence reversing the signs of its flux-linkage in accordance to its rotor positions [11]. The proposed FRDC machine purposely allocates its DC-field windings in the modulated stator slots in a way to imitate the FRPM configuration, and thus offering the same reversing flux-linkage pattern as the FRPM machines do. To achieve it, each of the stator pole is modulated as a 3-tooth per stator pole machine as indicated in Fig. 6.2. When the FRDC machine rotates its rotor from position 1 to 2, the polarities of its flux-linkages interchange accordingly, as shown in Fig. 6.3. With the bipolar flux-linkage characteristic, the proposed FRDC generator enjoys the higher power density as compared with its unipolar flux-linkage counterpart [12].

6.3.2   Selection of the Multiphase Topologies

The no-load electromotive force (EMF) waveforms may exhibit infinite patterns in reality, while all these waveforms can be roughly classified into two major categories, namely the sinusoidal-like waveform and the trapezoidal-like waveform. Meanwhile, the no-load EMF patterns can be regulated by the machine structure as well as the pole-pair arrangement.

When the no-load EMF e_s and phase current i_s are in phase with sinusoidal patterns, the average output power P_s can be expressed as

$$P_s = \frac{1}{\pi} \int_0^\pi e_s(\omega t) i_s(\omega t) \mathrm{d}(\omega t) = \frac{1}{2} E_s I_s \qquad (6.6)$$

where E_s and I_s are the amplitudes of the no-load EMF and phase current with sinusoidal-like waveforms, respectively. Similarly, when the no-load EMF e_t and

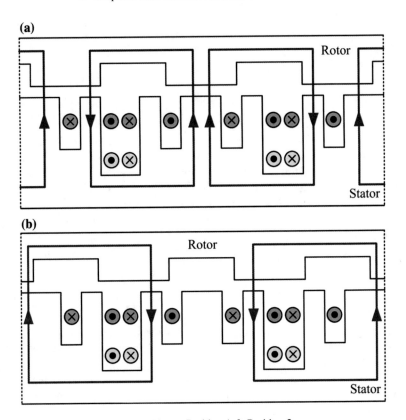

Fig. 6.3 Flux pattern of FRDC machine. **a** Position 1. **b** Position 2

phase current i_t are in phase with trapezoidal patterns, the average output power P_t can be expressed as

$$P_t = \frac{1}{\pi}\int_0^\pi e_t(\omega t)i_t(\omega t)\mathrm{d}(\omega t) = \frac{1}{3}\left(2k_t + \frac{6}{m}\right)E_t I_t \tag{6.7}$$

where k_t is the inclining slopes of the trapezoidal waveform, E_t and I_t are the amplitudes of the no-load EMF and phase current with trapezoidal-like waveforms, respectively. By combining the Eqs. (6.6) and (6.7), the power ratio of the sinusoidal to trapezoidal types can be further deduced as

$$\frac{P_s}{P_t} = \left(\frac{3m}{4k_t m + 12}\right)\left(\frac{E_s I_s}{E_t I_t}\right) \tag{6.8}$$

The rms no-load EMF ratio of the sinusoidal to trapezoidal patterns can be described as

Table 6.2 Power ratio of trapezoidal-type to sinusoidal-type

m	1	2	3	4	5	6
P_t/P_s	0.64	0.90	1.10	1.27	1.42	1.56

$$\frac{E_s}{E_t} = \frac{\pi}{2}\left(\frac{k_t m + 2}{m}\right) \tag{6.9}$$

Similarly, the rms phase current ratio of the sinusoidal to trapezoidal patterns can be described as

$$\frac{I_s}{I_t} = \sqrt{\frac{4k_t m + 12}{3m}} \tag{6.10}$$

By combining Eqs. (6.8), (6.9) and (6.10), the power ratio can be further deduced as

$$\frac{P_t}{P_s} = \frac{4m}{(k_t m + 2)\pi}\sqrt{\frac{k_t m + 3}{3m}} \tag{6.11}$$

In the ideal case, the inclining slopes of the trapezoidal waveform should be infinite, leading to achieving the square patterns, i.e., $k_t = 0$. By taking this criterion into consideration, the power ratios at different phases are calculated and categorized in Table 6.2. According to the results, it can be shown that starting from $m = 3$, the trapezoidal-like, or square-like waveform generators can provide higher output powers than the sinusoidal-like waveform counterparts do. This situation becomes even more obvious when it comes to higher phases. Therefore, the FRDC generator, which inherits the trapezoidal-like no-load EMF waveform, is more preferable to adopt the multi-phase topologies.

6.4 Generator Performance Analysis

6.4.1 Electromagnetic Field Analysis

The electromagnetic field analysis has been one of the most accurate and convenient tools to study electric machine performances for many years. In general, it can be classified into two major categories, namely the analytical field calculation and the numerical field calculation. To perform the machine modeling, the electromagnetic field equations are governed by

$$
\begin{cases}
\Omega : \dfrac{\partial}{\partial x}\left(v\dfrac{\partial A}{\partial x}\right) + \dfrac{\partial}{\partial y}\left(v\dfrac{\partial A}{\partial y}\right) = -(J_z + J_f) \\[2mm]
A|_S = 0
\end{cases}
\tag{6.12}
$$

where Ω is the field solution region, A and J_z the z-direction components of vector potential and current density, respectively, J_f the equivalent current density of the excitation field, S the Dirichlet boundary and v the reluctivity. Meanwhile, the equivalent circuit equation during generation is described as

$$
(R + R_L)i_s + (L + L_L)\frac{di}{dt} - \frac{l}{s}\iint_\Omega \frac{\partial A}{\partial t}\, d\Omega = 0
\tag{6.13}
$$

where R is the winding resistance, R_L the load resistance, L the end winding inductance, L_L the load inductance, l the axial length and s the conductor area of each turn of phase winding. The FEM is applied to analyze the machine performances of the proposed FRDC generator and the JMAG-Designer is employed as the magnetic solver to perform the FEM. The magnetic field distribution of the proposed generator at no-load condition is shown in Fig. 6.4. The result shows that

Fig. 6.4 Magnetic field distribution of the proposed FRDC machine

the flux distributions of the proposed FRDC generator are well balanced and also align with the theoretical expectations as shown in Fig. 6.3.

6.4.2 Pole-Arc Improved Design

In order to avoid the magnetic saturation, the values of the modulated stator tooth width β_s and the rotor pole-arc β_r should be well-tuned and confirmed with a particular value. Since the middle modulated stator tooth converges two balanced fluxes from both sides, its tooth width should be twice larger than that of the sided-tooth. At the primitive stage, the rotor pole-arc β_r is set equal to the value of β_s, i.e., $\beta_r = \beta_s$, as shown in Fig. 6.5a. Meanwhile, the β_r can be fine-tuned by improving the pole-arc ratio as described with the following expression: $\beta_{r\_imp} = n\beta_s$, as shown in Fig. 6.5b. The variations of the no-load EMF according to the pole-arc ratio n are shown in Fig. 6.6. As previously discussed, the trapezoidal-like no-load EMF waveform with well-balanced pattern is more desirable for the wind power generation, and hence the pole-arc ratio should be selected in the range of $n = 1.4–1.6$.

To confirm the pro-arc ratio with the optimal performance, the cogging torque is also studied and its waveforms under different pole-arc ratios are shown in Fig. 6.7. In the case when $\beta_r = \beta_s$, i.e., $n = 1.0$, the peak value of the cogging torque is approximately 29.2 N m. When the pole-arc ratio is set to be $n = 1.5$, the cogging torque will then be minimized to 11.3 N m, which is only 6.28% of its rated torque. This particular cogging torque value is within the acceptable range, as compared with the commonly employed PM candidates [12]. In general, the lower the cogging torque value, the better the machine performances. Hence, it can be confirmed

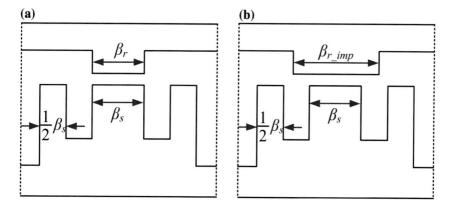

Fig. 6.5 Topologies of the modulated stator and rotor poles. **a** Primitive β_r. **b** Improved β_{r_imp}

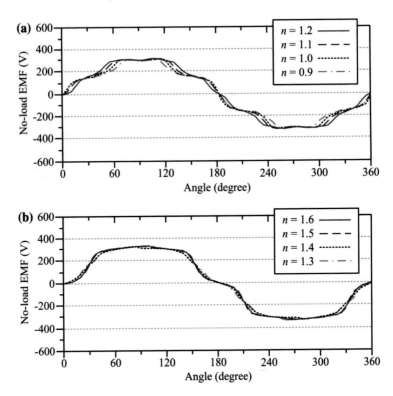

Fig. 6.6 No-load EMF waveforms under different pole-arc ratios. **a** n = 0.9–1.2 **b** n = 1.3–1.6

that the proposed generator has been improved with the pole-arc ratio at $\beta_{r\_imp} = 1.5\beta_s$.

6.4.3   Proposed FRDC Performance Analysis

The DC flux-linkage waveforms of the proposed FRDC machine at no-load condition are shown in Fig. 6.8. The result shows the proposed machine obtains the bipolar DC flux-linkage patterns, which behaves similarly as the traditional FRPM machine does. In addition, the simulation results show the proposed generator obtains the four-phase flux-linkage pattern without noticeable distortion. Hence, these results verify the proposed machine structure and the winding arrangement are correct.

The no-load EMF waveforms of the proposed FRDC machine under the operating speed of 900 rpm are shown in Fig. 6.9. It can be shown that the no-load EMF waveforms are well-balanced with the trapezoidal-like pattern. Furthermore, the no-load EMF waveforms at various speeds under two conditions, namely

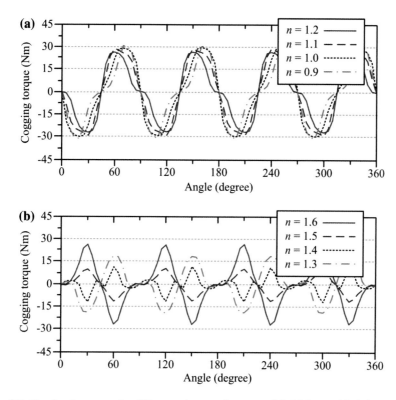

Fig. 6.7 Cogging torques under different pole-arc ratios. **a** $n = 0.9$–1.2 **b** $n = 1.3$–1.6

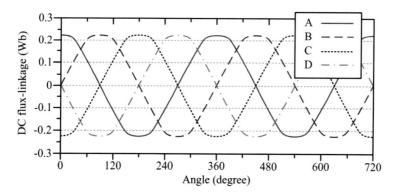

Fig. 6.8 DC flux-linkage waveforms under no-load condition

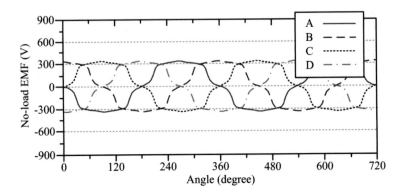

Fig. 6.9 No-load EMF waveforms under operating speed of 900 rpm

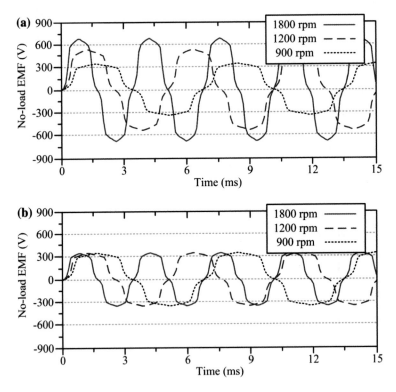

Fig. 6.10 No-load EMF waveforms under various operating speeds. **a** Without flux regulation. **b** With flux regulation

without and with flux regulations are shown in Fig. 6.10a and b, respectively. To illustrate the no-load EMF waveforms under different operating speeds as well as different frequencies, the waveforms in Fig. 6.10 are purposely plotted against time, instead of angle. Undoubtedly, without any DC-field flux regulations, the generated voltages vary along with different operating speeds. These varying output voltages are not desirable for the wind power generation because these fluctuating voltages may exceed the system threshold value, and hence damaging the whole system. On the other hand, with the support of the DC-field control, the output voltages can be maintained in a certain value over the wide-speed ranges.

6.4.4 Comparison with FRPM Machines

To manifest the merits of the proposed magnetless four-phase FRDC machine, the well-developed FRPM machines, namely the three-phase FRPM and four-phase FRPM machines are included for comparisons. To offer the fair comparison, all the generators obtain the same machine dimensions, namely the outside diameters, stack lengths, airgap lengths and winding fill factors. The corresponding key design data of all machines are listed in Table 6.3.

With the application of the FEM, the performances of all machines are calculated and summarized in Table 6.4. In particular, the power densities of the three-phase FRPM, the four-phase FRPM, and the proposed FRDC generators are found to be 1.98, 2.10 and 0.71 MW/m^3, respectively. As discussed in Sect. 6.4.3, in the higher number of armature phase situations, the trapezoidal-like no-load EMF generators can result higher output powers than that of the sinusoidal-like counterparts. Hence,

Table 6.3 Key design data of the proposed FR machines

Items	FRPM		FRDC
Rotor outside diameter (mm)	448.0		448.0
Rotor inside diameter (mm)	353.0		353.0
Stator outside diameter (mm)	350.0		350.0
Stator inside diameter (mm)	76.0		76.0
No. of rotor poles	8	10	10
No. of actual stator poles	6	8	8
No. of pole-pairs on stator pole	1	1	1
No. of modulated stator tooth	N/A		3
Rotor pole arc (°)	22.0	18.0	18.0
Stator (modulated) pole arc (°)	48.0	36.0	12.0
Airgap length (mm)	1.5		1.5
Stack length (mm)	280.0		280.0
No. of armature phases	3	4	4
No. of turns per armature coil	1008	912	456

Table 6.4 Comparisons between the FRDC and FRPM generators

Items	FRPM		FRDC
No. of armature phases	3	4	4
Power (kW)	58	64	22
Power density (MW/m³)	1.98	2.10	0.71
Flux controllability	Low	Low	High
Material cost (USD)	1245	1398	308
Cost-effectiveness (W/USD)	46.6	45.8	71.4

the FR generators, which inherit the trapezoidal-like no-load EMF waveforms with the four-phase topology, can offer higher power density than that of the three-phase one. The FEM results verify the theoretical expectation in which the four-phase FRPM generator can provide higher output power that of the three-phase one, and hence illustrating the merit of the multi-phase FR generators.

By employing the high-power-density PM materials for field excitations, the FRPM machines can offer relatively higher power densities, as compared with the proposed magnetless FRDC counterpart. However, the PM generators suffer from the uncontrollable flux problem, and hence the PM candidates may overcharge the battery or even damage the wind power system at some strong wind situations. In addition, the supply of the PM materials is limited and floating, leading to the raised material costs.

The proposed magnetless generator undoubtedly suffers from the shortcomings of lower power density, yet a more comprehensive comparison among generators should consider flux controllability and cost-effectiveness. By utilizing the DC-field winding for excitation, the proposed FRDC machine can eliminate the high-cost PM materials, and thus accomplishing the desired performance regarding its high flux controllability and effective cost, as compared with its counterparts. Hence, it can be expected that the proposed magnetless candidate may demonstrate promising potential in the wind power generation industry.

6.5 Experimental Verifications

To verify the proposed idea, an experimental setup for the proposed FRDC generator is established and as shown in Fig. 6.11. In order to perform sensible and practical experiments in the laboratory, the power level of the prototype is scaled down on purpose. Hence, the measured results are reduced proportionally, as compared with the simulated one.

The measured no-load EMF waveforms of the proposed FRDC generator under the operating speed of 900 rpm are shown in Fig. 6.12. As illustrated, the measured waveforms are slightly different from the simulated results as shown in Fig. 6.9, and the differences are typically caused by the manufacturing imperfection. Meanwhile, the rms values of the waveforms still well agree with the theoretical

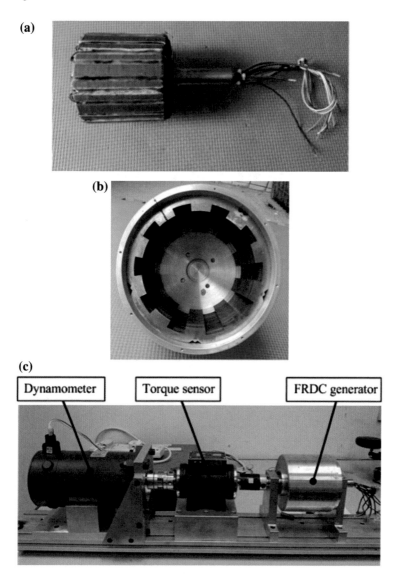

Fig. 6.11 Experimental setup. **a** FRDC stator segment. **b** FRDC rotor segment. **c** Test bed

ones, and hence the discrepancies are believed to be acceptable. In addition, the measured no-load EMF waveforms under the higher speed of 1800 rpm, without and with the flux regulating controls, are shown in Fig. 6.13a and b, respectively. By purposely regulating the DC-field excitation, the proposed wind power generation system can maintain its generated voltage at the same level over the wide-speed ranges, as aligning with the theoretical results in Fig. 6.10.

Fig. 6.12 Measured no-load
EMF waveforms at 900 rpm

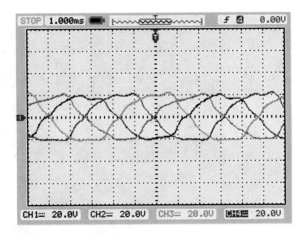

Fig. 6.13 Measured no-load
EMF waveforms at 1800 rpm.
a Without flux regulation.
b With flux regulation

(a)

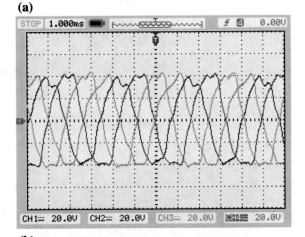

(b)

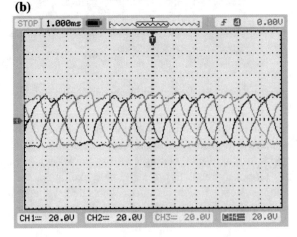

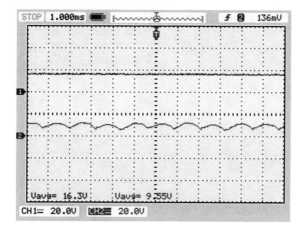

Fig. 6.14 Measured DC voltage (Trace 1) and load current (Trace 2)

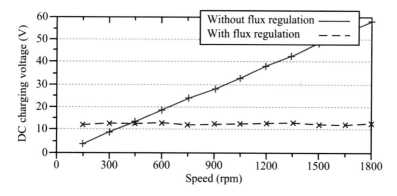

Fig. 6.15 Measured voltage characteristics versus speed at no load

Through the AC-DC rectification, the generated voltage is rectified to give out the DC charging voltage and load current, as shown in Fig. 6.14. The rectified dc output can be connected to the battery as the energy storage, and the stored energy can then be transferred to the power grid after the DC-AC conversion performed by the inverter. Furthermore, the corresponding variations of the measured DC voltage with respect to the operating speed at no-load, without and with flux regulations, are shown in Fig. 6.15. The results confirm that the wind power generation system equipped with the proposed FRDC machine can utilize its flux-controlled capability to keep the DC charging voltages at the certain level, and hence extending the battery life cycle and protecting the whole system. In addition, the DC charging performances of the proposed system under the load conditions are also analyzed. The corresponding variations of the measured DC charging voltage characteristics under the operating speed of 900 rpm with respect to the load current, without and

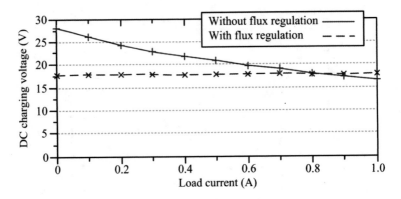

Fig. 6.16 Measured charging voltage characteristics versus load current

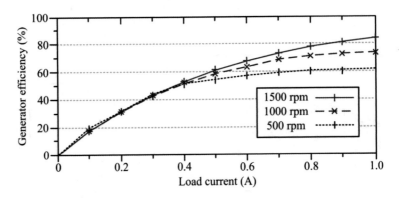

Fig. 6.17 Measured efficiencies versus load current at different speeds

with flux regulations, are shown in Fig. 6.16. Not surprisingly, the DC charging voltages can be once again maintained at the pre-assigned level. Hence, the results further exemplify the proposed FRDC generator can behave as a constant-output-voltage generator at the wide ranges of wind speeds and load currents, and these characteristics are particularly favorable for the modern wind power generation applications.

Finally, the system efficiencies are measured under different operating speeds with different load currents, as shown in Fig. 6.17. It can be found that the efficiencies can be kept at high standards over a wide range of speed with different load currents. To be specific, the efficiency under the speed of 1500 rpm with load current of 1.0 A is around 83.5%. The proposed FRDC machine can achieve comparable efficiencies as compared with the profound PM counterparts do, whose efficiencies are around 80% [6].

6.6 Summary

This chapter introduces a new four-phase FRDC generator for the wind power generation applications. The key of the proposed machine is to purposely modulate the stator pole in a way to imitate the FR patterns, and hence allowing the machine to behave similarly as the FRPM machine does. By utilizing the controllable DC-field winding, the proposed FRDC generator can control its airgap flux densities, and thus achieving the constant-output-voltage charging characteristics under different conditions. With the magnetless structure, the proposed FRDC machine enjoys the predominance of cost-advantage as compared with its PM counterparts. Both the FEM analysis and experimental verifications have confirmed the feasibility of the proposed wind power generation system.

References

1. J.P. Barton, D.G. Infield, Energy storage and its use with intermittent renewable energy. IEEE Trans. Energy Convers. **19**(2), 431–438 (2004)
2. F.M. Hughes, O. Anaya-Lara, N. Jenkins, G. Strbac, Control of DFIG-based wind generation for power network support. IEEE Trans. Power Syst. **20**(4), 1958–1966 (2005)
3. M. Chinchilla, S. Arnaltes, J.C. Burgos, Control of permanent-magnet generators applied to variable-speed wind-energy systems connected to the grid. IEEE Trans. Energy Convers. **21**(1), 130–135 (2006)
4. G.O. Cimuca, C. Saudemont, B. Robyns, M.M. Radulescu, Control and performance evaluation of a flywheel energy-storage system associated to a variable-speed wind generator. IEEE Trans. Industr. Electron. **53**(4), 1074–1085 (2006)
5. A. Grauers, Efficiency of three wind energy generator systems. IEEE Trans. Energy Convers. **11**(3), 650–657 (1996)
6. Y. Fan, K.T. Chau, M. Cheng, A new three-phase doubly salient permanent magnet machine for wind power generation. IEEE Trans. Ind. Appl. **42**(1), 53–60 (2006)
7. A.J. Squarezi Filho, E.R. Filho, Model-based predictive control applied to the doubly fed induction generator direct power control. IEEE Trans. Sustain. Energy **3**(3), 398–406 (2012)
8. J.A. Krizan, S.D. Sudhoff, A design model for salient permanent-magnet machines with investigation of saliency and wide-speed-range performance. IEEE Trans. Energy Convers. **28**(1), 95–105 (2012)
9. C.X. Wang, I. Boldea, S.A. Nasar, Characterization of three phase flux reversal machine as an automotive generator. IEEE Trans. Energy Convers. **16**(1), 74–80 (2001)
10. C.H.T. Lee, K.T. Chau, C. Liu, C.C. Chan, Overview of magnetless brushless machines. IET Electr. Power Appl. (to appear)
11. R.P. Deodhar, S. Andersson, I. Boldea, T.J.E. Miller, The flux-reversal machine: a new brushless doubly-salient permanent-magnet machine. IEEE Trans. Industr. Electron. **33**(4), 925–934 (1997)
12. M. Cheng, W. Hua, J. Zhang, W. Zhao, Overview of stator-permanent magnet brushless machines. IEEE Trans. Industr. Electron. **58**(11), 5087–5101 (2011)

Chapter 7
Proposed Dual-Mode Machine for Wind Power Harvesting

7.1 Introduction

Because of the increasing concerns on the energy utilization and the environmental protection, the wind power has gained lots of attentions recently [1, 2]. The wind power harvesting applications generally can be classified as two big families, namely the constant-speed constant-frequency (CSCF) wind power generation and the variable-speed constant-frequency (VSCF) wind power generation [3, 4]. Without installation of power converters, the CSCF type takes the absolute merit of compact system architecture as well as simplicity. Yet, its operating speed has to be maintained in a limited range so that there is no correlation with the wind speed. Therefore, it suffers from the problems of lower efficiency and higher mechanical losses.

Due to the advancement of the power electronics, the VSCF system can be employed efficiently. Unlike the CSCF system, the VSCF system instead can absorb the maximum wind power based on the variable wind speed [5, 6]. Undoubtedly, the VSCF one can provide better efficiency than the CSCF does. The electric machines, as the key part of the VSCF system, have to provide some distinguished features, such as high efficiency level, satisfactory power density, good controllability, wide operating range, maintenance-free and fault-tolerant operations [7, 8]. The doubly salient permanent-magnet (DSPM) machines that are able to fulfill most of the goals have drawn many attentions in the past few decades. Nevertheless, with the installation of the PM materials, the DSPM machines unavoidably face the shortcomings of high construction cost as well as the uncontrollable PM flux [9, 10]. To handle these fundamental demerits, the magnetless machines, namely the switched reluctance (SR) [11] and the doubly salient DC-field (DSDC) machines have become a hot research topic recently. Upon the employment of the independent DC-field excitation, the DSDC machine is able

© Springer Nature Singapore Pte Ltd. 2018
C. H. T. Lee, *Design, Analysis and Application of Magnetless Doubly Salient Machines*, Springer Theses,
https://doi.org/10.1007/978-981-10-7077-8_7

to control its flux density easily so that this type of machine is highly favorable for the low-torque high-speed operation. On the other hand, the multi-tooth flux-switching DC-field (FSDC) machine, which provides the bipolar flux-linkage, is instead suitable for high-torque low-speed operation. Even the operating speed in the wind power harvesting is confined in a particular range, such that some of the developed machines may satisfy all its requirements accordingly, the researches on the new design concepts with better flexibility and stability is still admirable.

This chapter targets to combine the design philosophies of the FSDC and DSDC machines, in order to create a new dual-mode FS-DSDC machine, purposely for the wind power harvesting industries. Upon the regulation of different winding architectures, the proposed machine behaves similarly as its corresponding precedents do. Therefore, the proposed machine is able to operate based on two separated modes, namely the FSDC mode for low-speed operation and the DSDC mode for high-speed operation. The finite element method (FEM) is adopted for machine performance analysis, while the experimental prototype is also developed to verify the proposed concepts.

7.2 System Configuration and Speed Consideration

Figure 7.1 shows the proposed architecture of the wide operating range wind power harvesting system that contains the key components as follows: (i) a gearbox-installed wind turbine for wind power adsorption, (ii) the proposed FS-DSDC machine for electromechanical energy conversion, (iii) the power electronic relay module for winding connection regulation, (iv) two three-phase full-bridge for AC-DC conversion, (v) two batteries for captured energy storage, (vi) a buck converters based on the DC-field controller signal for rectified dc voltage regulation and (vii) an inverter for DC-AC conversion.

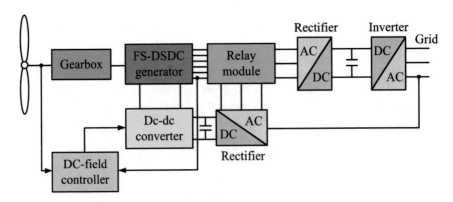

Fig. 7.1 Proposed wind power harvesting architecture

Based on the discussion of Betz's theory, the peak output power can be captured within a limited wind speed range [3]. Normally, the operating speed of the wind generator should never exceed 1000 rpm at the normal scenario, while it will be beyond 1000 rpm at the strong wind scenario.

7.3   Proposed Dual-Mode FS-DSDC Machine

7.3.1   Proposed Machine Topology

The structure of the proposed dual-mode FS-DSDC machine, which is adopted as the proposed wind power harvesting application is shown in Fig. 7.2a. The proposed FS-DSDC machine purposely integrate the design equations of FSDC and DSDC machines so that it can be operated with two different modes, namely the FSDC mode and the DSDC mode. The design equations for the FSDC and the DSDC machines are governed by Eqs. (7.1) and (7.2), respectively as shown below

$$\begin{cases} N_{sp} = 2mi \\ N_{se} = N_{sp}N_{st} \quad (N_{st} = 2, 4\ldots) \\ N_r = N_{se} - N_{sp} \pm 2i \end{cases} \tag{7.1}$$

$$\begin{cases} N'_{sp} = 2m'j \\ N'_{se} = N'_{sp}N'_{st} \\ N'_r = N'_{se} \mp 2k \end{cases} \tag{7.2}$$

where m, N_{sp}, N_{st}, N_{se}, and N_r are the number of armature phases, the stator pole, the stator teeth, the equivalent stator pole and the rotor pole for FSDC machine, respectively; m', N'_{sp}, N'_{st}, N'_{se}, and N'_r are the armature phases, the stator pole, the stator teeth, the equivalent stator pole and the rotor pole for DSDC machine, respectively; i, k and j are any integers. To couple the specific design philosopies between the two machine types, the number of equivalent stator poles of the two machines, i.e., N_{se} and N'_{se} are set equal such that the following deduction can be formed accordingly

$$(mi)N_{st} = (m'j)N'_{st} \tag{7.3}$$

Meanwhile, the derived equation provides infinitely many solutions and to minimize the degree of freedom, i and j are purposely set to be equal so that the equation can be further derived as

$$\frac{N_{st}}{N'_{st}} = \frac{m'}{m} \quad (m \neq m') \tag{7.4}$$

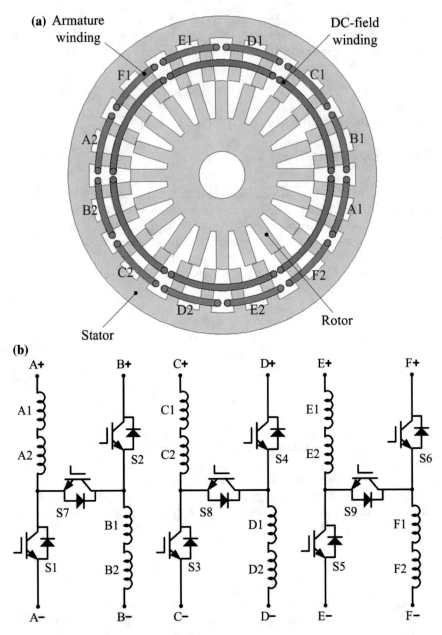

Fig. 7.2 Proposed dual-mode FS-DSDC machine. **a** Proposed machine structure. **b** Proposed winding connection configuration

Table 7.1 Basic design combinations of the FS-DSDC machine

m	m'	i	j	k	N_{st}	N'_{st}	$N_{se} = N'_{se}$	$N_r = N'_r$
3	6	1	1	2	2	1	12	8
3	6	1	1	2	4	2	24	20
4	8	1	1	3	2	1	16	10

Based on the same concept, the number of the rotor poles of two machines, i.e., N_r and N'_r are also set to be equal. With the implementation of Eq. (7.4), the relationship can be derived as

$$k = (m - 1)i \tag{7.5}$$

Based on Eqs. (7.1–7.5), the basic design combinations of the FS-DSDC machine can be deduced and classified in Table 7.1. To ease the control algorithm, the three-phase and six-phase structures, i.e., $m = 3$ and $m' = 6$ are selected. In the meantime, to offer the relatively lower speed range, say from 0 to 1000 rpm at normal wind scenario, the multi-tooth structure, i.e., $N_{st} = 4$ and $N'_{st} = 2$ is more suitable as compared with its single-tooth counterpart, i.e., $N_{st} = 2$ and $N'_{st} = 1$. By taking these discussions into considerations, the combinations of $N_{se} = 24$ and $N_r = 20$, as the proposed topology for the FS-DSDC machine are formed.

As referring to the conventional generators, the proposed one contains the controllable DC-field winding such that it can offer better flexibility for various situations. Even the voltage regulation can be achieved with the help of the independent converter modules, the proposed generator can instead employ its built-in flux regulation capability to cope with all different missions as the all-in-one system. Furthermore, the proposed generator can be operated based on two different modes. Therefore, it can be expected that the proposed generator can provide better fault-tolerability as compared with its conventional counterparts.

7.3.2   Dual-Mode Operating Principle

The two armature windings both employ the concentrated winding arrangements, so that both of them can be easily developed. Every armature winding loop is independently connected such that all of them can be separately conducted. Hence, various connecting architectures can be accomplished based on the support of the power electronic relay module.

The proposed winding connection configuration for the proposed machine is shown in Fig. 7.2b. For the FSDC mode, the switches S1, S2, S3, S4, S5, and S6 are off, while S7, S8, and S9 are on, so that the armature windings are connected to offer the following settlement: A1, A2, B1 and B2 are connected in series; C1, C2, D1 and D2 in series; E1, E2, F1 and F2 in series. Hence, the proposed machine can behave similarly as the three-phase FSDC machine does. On the other hand, for the

DSDC mode, the switches S1, S2, S3, S4, S5, and S6 are on, while S7, S8, and S9 are off, so that the armature windings are connected in a way similarly as a six-phase DSDC machine does.

The winding arrangement of the proposed machine at the FSDC mode behaves similarly as the two armature winding sets connecting in series, so that the generated voltage is higher than that from the DSDC mode does. Because the generated voltage changes along with the wind speed directly, the FSDC mode should be adopted for the low-speed situation, so that the generated voltage will never overshoot the threshold limit. On the other hand, the DSDC mode should be instead adopted for the high-speed situation.

With the support of the relay module, the three-phase armature winding at FSDC mode can be connected with one single three-phase full-bridge rectifier for AC-DC conversion to charge up the battery. Consequently, the energy can then be transferred to the grid via the inverter. In the meantime, the six-phase armature winding at DSDC mode can be switched to connect with two three-phase full-bridge rectifiers, where one of them is used for energy transmission to the grid while another one for battery charging to support the DC-field excitation. Even under the extremely strong wind situation, this proposed setup can guarantee the developed DC voltage will never overshoot the normal voltage limit. Furthermore, the additional DC-field energy storage can minimize the impact to the grid, so that the interference to the power grid can then be reduced.

7.4 Finite Element Method Analysis

In many years, the electromagnetic field analysis has always been one of the most accurate and convenient tools to perform the electric machine analysis. Basically, it can be categorized into two major subsidiaries, namely the analytical field calculation [12] and the numerical field calculation [13]. To model the machine performances, the electromagnetic field equation is described as

$$\begin{cases} \Omega : \frac{\partial}{\partial x}\left(v\frac{\partial A}{\partial x}\right) + \frac{\partial}{\partial y}\left(v\frac{\partial A}{\partial y}\right) = -(J_z + J_f) \\ A|_s = 0 \end{cases} \tag{7.6}$$

where Ω is the field solution region, A and J_z the z-direction components of vector potential and current density, respectively, J_f the equivalent current density of the excitation field, S the Dirichlet boundary and v the reluctivity. In addition, the equivalent circuit equation during generation is governed by:

$$(R + R_L)i_s + (L + L_L)\frac{di}{dt} - \frac{l}{s}\iint_{\Omega} \frac{\partial A}{\partial t} d\Omega = 0 \tag{7.7}$$

where R is the winding resistance, R_L the load resistance, L the end winding inductance, L_L the load inductance, l the axial length and s the conductor area of each turn of phase winding. The FEM is applied to analyze the machine performances of the proposed FS-DSDC machine and the JMAG-Designer is employed as the magnetic solver to perform the FEM. The magnetic field distribution of the proposed machine at no-load condition is shown in Fig. 7.3.

The airgap flux density waveform of the proposed machine at no-load condition is shown in Fig. 7.4. The results show the original flux of each stator pole is modulated into four portions in accordance with the number of teeth per stator pole. According to the corresponding winding configurations, the proposed machine at FSDC mode acts similarly as the four-toothed FSDC machine does, while at DSDC mode as the two-toothed DSDC machine does.

The DC flux-linkage waveforms at the FSDC mode and at the DSDC mode are shown in Fig. 7.5a, b, respectively. The simulation waveforms show that the proposed machine obtains the bipolar DC flux-linkage pattern at FSDC mode, while the unipolar DC flux-linkage pattern at DSDC mode. These results confirm that the pole-pair arrangements and the armature winding configurations of the proposed machine are correct. Hence, it can further verify the proposed design equations are correct.

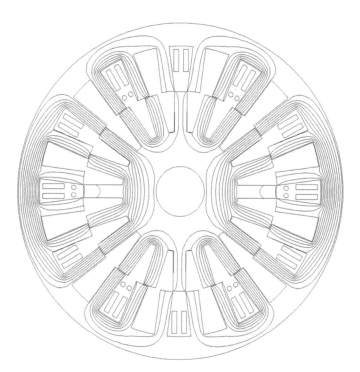

Fig. 7.3 Magnetic field distribution of the proposed FS-DSDC machine

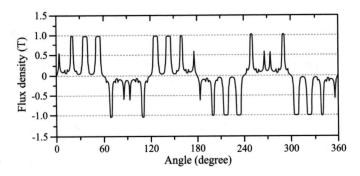

Fig. 7.4 Airgap flux density of the proposed FS-DSDC machine

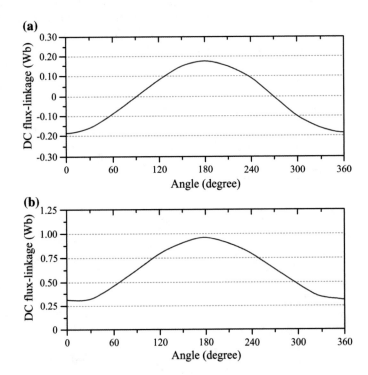

Fig. 7.5 DC flux-linkage waveforms. **a** FSDC mode. **b** DSDC mode

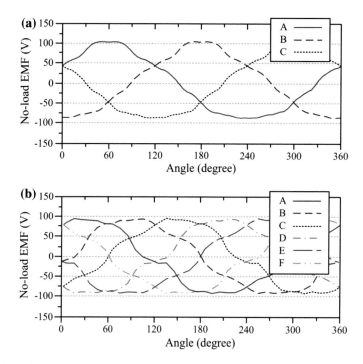

Fig. 7.6 No-load EMF waveforms. **a** At FSDC mode with speed of 300 rpm. **b** At DSDC mode with speed of 600 rpm

By performing FEM, the performances of the proposed wind power generation system can be analyzed thoroughly. The no-load electromotive force (EMF) waveforms of the FS-DSDC machine at FSDC mode under the operating speed of 300 rpm and at DSDC mode of 600 rpm are shown in Fig. 7.6a, b, respectively. It can be shown that the no-load EMF waveforms at FSDC mode are well balanced among three-phase pattern without significant distortion, while at DSDC mode are six-phase waveforms with symmetrical pattern instead. The magnitudes of the no-load EMF among each phase at FSDC mode is approximately the same as compared with those at DSDC mode. Hence, these results confirm that the FSDC mode should be adopted at the low-speed environment, while the DSDC mode at the high-speed environment.

In addition, the no-load EMF waveforms under the FDSC mode and DSDC mode at various speeds under two conditions, namely without and with flux regulations are shown in Figs. 7.7 and 7.8, respectively. Undoubtedly, without any DC-field flux regulations, both the generated voltages at two modes vary along with the varying speeds. Therefore, the output voltages may overcharge the battery and harm the whole wind power system. Meanwhile, the DC-field level can be tuned purposely in order to keep the output voltages as constant as possible over the wide-speed ranges.

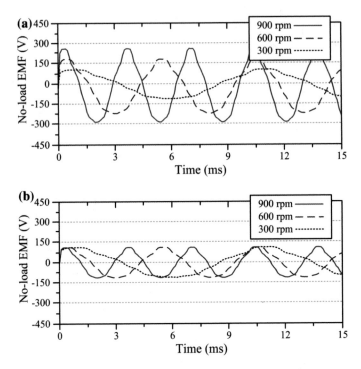

Fig. 7.7 No-load EMF waveforms at FSDC mode with various operating speeds. **a** Without flux regulation. **b** With flux regulation

7.5 Experimental Verifications

For verification, the experimental setup of the proposed wind power generator under the open-loop mode is established and as shown in Fig. 7.9. The corresponding key design data of the proposed FS-DSDC machine is listed in Table 7.2.

The measured no-load EMF waveforms of the proposed FS-DSDC machine at FSDC mode and DSDC mode under the corresponding conditions are shown in Fig. 7.10. As shown, the measured waveforms are slightly different from the simulated ones as shown in Fig. 7.6a, b, and the differences are generally caused by the manufacturing imperfection. Meanwhile, the experimental results are still within the normal range and the discrepancies are expected to be acceptable. In addition, the measured no-load EMF waveforms at FSDC mode under the higher speed of 900 rpm and at DSDC mode of 1800 rpm, without and with the DC-field flux regulations are shown in Fig. 7.11. Since all the no-load EMF waveforms of the six-phase windings at DSDC mode are well balanced and symmetrical, to have a better illustration, only the phase A, C, and E are shown purposely. With the flux regulation, the magnitudes of the no-load EMF waveforms at higher speed ranges can be kept at the same level as compared with those at the lower speed ranges.

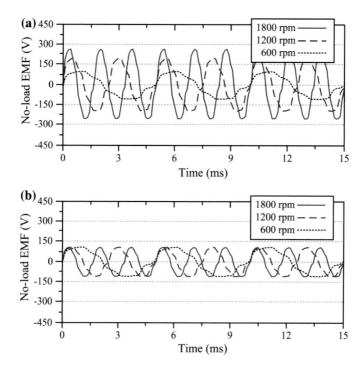

Fig. 7.8 No-load EMF waveforms at DSDC mode with various operating speeds. **a** Without flux regulation. **b** With flux regulation

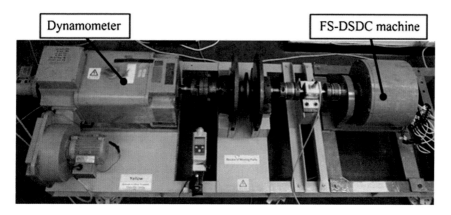

Fig. 7.9 Experimental setup of the wind power generation system

Table 7.2 Key data of the proposed FS-DSDC machine

Items	FS-DSDC
Stator outside diameter (mm)	270.0
Stator inside diameter (mm)	161.2
Rotor outside diameter (mm)	160.0
Rotor inside diameter (mm)	40.0
No. of equivalent stator poles	24
No. of rotor poles	20
Stator pole arc (°)	7.5
Rotor pole arc (°)	7.5
Stator pole height (mm)	36
Rotor pole height (mm)	32
Airgap length (mm)	0.6
Stack length (mm)	80.0
No. of turns per armature coil	100

These measured waveforms verify the expected performances by the simulation waveforms in Figs. 7.7 and 7.8, and hence confirming the proposed wind power generation system is able to maintain the generated voltages to become as constant as possible over a wide range of operating speeds, even under the strong wind situation.

Through the three-phase rectifications, the simulated and the measured DC charging voltages at the FSDC and at DSDC modes are shown in Fig. 7.12. The single output at FSDC mode can be connected to the grid for energy transmission, while the two outputs at DSDC mode can be connected to the grid and also to the DC-field energy storage device. In addition, the corresponding simulated and measured DC charging voltage characteristics at FSDC mode and at DSDC mode, without and with the flux regulations, with respect to the operating speed at no-load are shown in Fig. 7.13, where the measured results well agree with the simulated ones. The results verify that the proposed wind power generation system can utilize the flux regulating ability to maintain the DC charging voltages at the certain level, and hence protecting the whole system. Meanwhile, the charging performances of the proposed system under the load conditions are also studied. The corresponding simulated and measured DC charging voltage characteristics at FSDC mode under the operating speed of 900 rpm and at DSDC mode of 1800 rpm, without and with the flux regulations, with respect to the load current are shown in Fig. 7.14. Once again, by regulating the DC-field winding, the DC charging voltages can be kept at the specific level. Therefore, the results further verify the proposed wind power generation system can perform as a constant-voltage generator at the wide range of wind speeds and load currents.

Finally, the efficiencies of the proposed machine at two modes as the function of speeds have been measured and shown in Fig. 7.15. Particularly, the efficiencies of the proposed machine at FRDC mode and at DSDC mode can achieve approximately 74 and 72%, respectively. It can be found that the proposed machine at both modes can achieve comparable efficiencies as compared with the commonly

Fig. 7.10 Measured no-load
EMF waveforms. **a** FSDC
mode. **b** Phase A, C, and E at
DSDC mode. **c** Phase B, D,
and F at DSDC mode

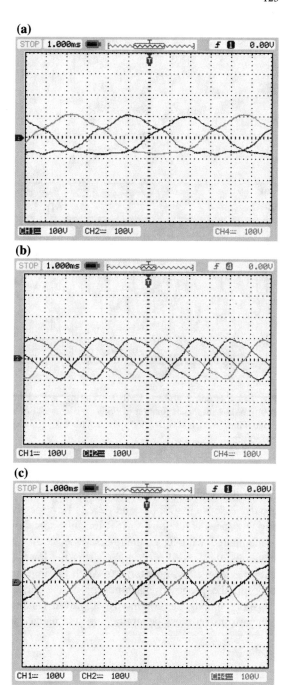

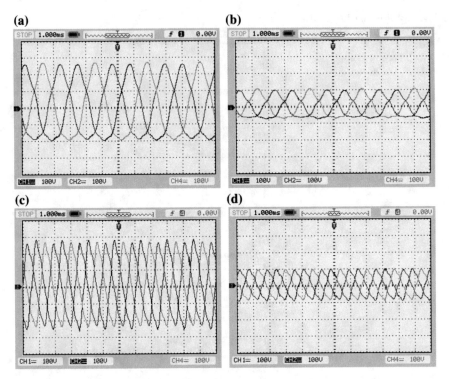

Fig. 7.11 Measured no-load EMF waveforms. **a** Without flux regulation of FSDC mode at 900 rpm. **b** With flux regulation of FSDC mode at 900 rpm. **c** Without flux regulation of DSDC mode (Phase A, C, and E) at 1800 rpm. **d** With flux regulation of DSDC mode (Phase A, C, and E) at 1800 rpm

employed machines do. To be specific, the efficiency of the magnetless SR generator, the doubly-fed induction generator (DFIG), and the DSPM generator are around 70% [11], 80% [4], and 82% [14], respectively. Despite of the flexibility in variable speed conditions, the SR machine somehow has relatively lower efficiency as compared with its counterparts. Due to the installation of the high-cost PM materials, the DSPM type suffers from the lowest cost-effectiveness as compared with the magnetless counterparts. With the implementation of the mode-changing concept, the proposed machine enjoys the highest flexibility in variable speed situations among the commonly employed candidates. For better illustration, the comparisons on these generators are summarized in Table 7.3.

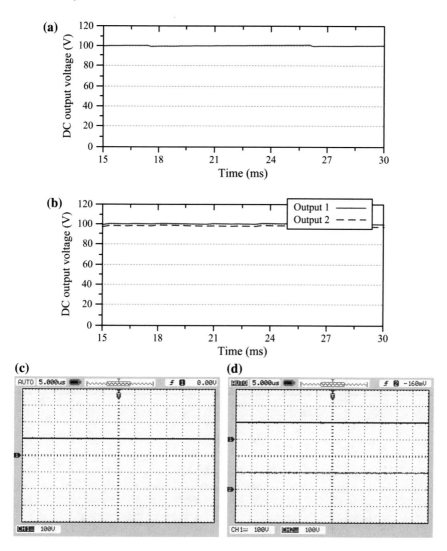

Fig. 7.12 DC charging voltages. **a** Simulated result of FSDC mode. **b** Simulated result of DSDC mode. **c** Measured result of FSDC mode. **d** Measured result of DSDC mode

7.6 Summary

This chapter introduces a new dual-mode FS-DSDC machine for wind power harvesting applications. The key of the proposed machine is to purposely incorporate two machine design philosophies together, and hence allowing the machine to offer higher flexibility and stability to cater different possible situations. Based on the independent DC-field excitations, the proposed machine can produce the

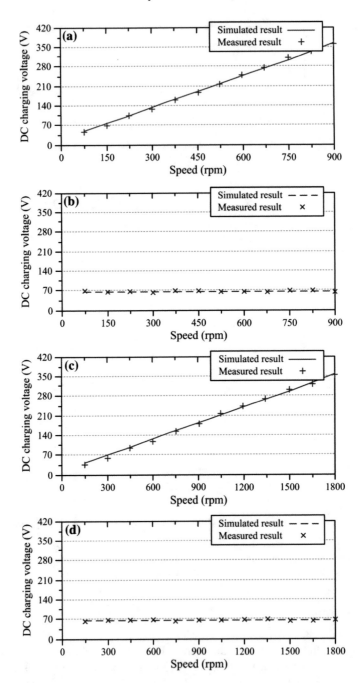

Fig. 7.13 Measured charging voltage characteristics versus speed at no load. **a** FSDC mode—without flux regulations. **b** FSDC mode—with flux regulations. **c** DSDC mode—without flux regulations. **d** DSDC mode—with flux regulations

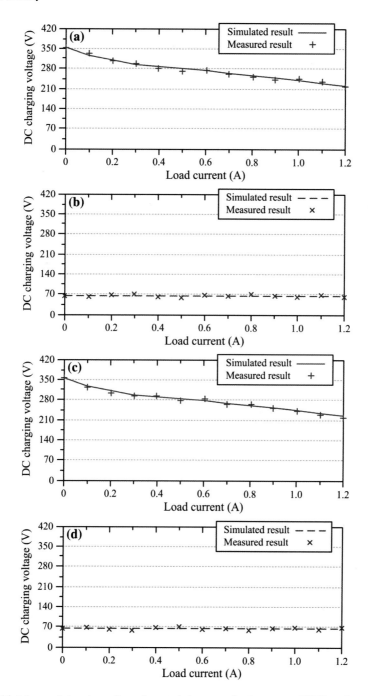

Fig. 7.14 Measured charging voltage characteristics versus load current. **a** FSDC mode—without flux regulations. **b** FSDC mode—with flux regulations. **c** DSDC mode—without flux regulations. **d** DSDC mode—with flux regulations

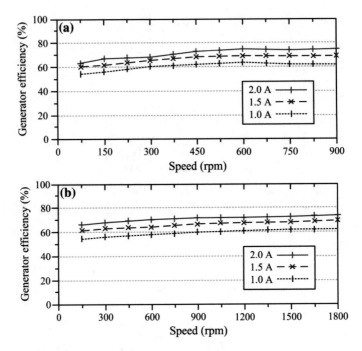

Fig. 7.15 Measured efficiencies versus speed at different loads. **a** FSDC mode. **b** DSDC mode

Table 7.3 Comparisons among different common generators

Items	SR	DFIG	DSPM	FS-DSDC
Efficiency	Low	High	High	Moderate
Cost-effectiveness	High	High	Low	High
Flexibility	Moderate	High	Moderate	Very high

controllable airgap flux densities, and thus accomplishing the constant-voltage charging characteristics over a wide-speed range under different load currents. Without implementation of any PM material, the proposed magnetless machine takes the absolute advantage of cost-benefit as compared with its PM counterparts. Both the FEM simulations and experimental results confirm the validity of the proposed wind power harvesting system.

References

1. G.W. Ault, K.R.W. Bell, S.J. Galloway, Calculation of economic transmission connection capacity for wind power generation. IET Renew. Power Gener. **1**(1), 61–69 (2007)
2. P.J. Luickx, E.D. Delarue, W.D. D'haeseleer, Effect of the generation mix on wind power introduction. IET Renew. Power Gener. **3**(3), 267–278 (2009)
3. L. Jian, K.T. Chau, J.Z. Jiang, A magnetic-geared outer-rotor permanent-magnet brushless machine for wind power generation. IEEE Trans. Ind. Appl. **45**(3), 954–962 (2009)
4. Y. Han, S. Kim, J.I. Ha, W.J. Lee, A doubly fed induction generator controlled in single-sided grid connection for wind turbine. IEEE Trans. Energy Convers. **28**(2), 413–424 (2013)
5. F.D. Kanellos, N.D. Hatziargyrious, Control of variable speed wind turbines equipped with synchronous or doubly fed induction generators supplying islanded power systems. IET Renew. Power Gener. **3**(1), 96–108 (2009)
6. C.H.T. Lee, K.T. Chau, C. Liu, Design and analysis of a cost-effective magnetless multi-phase flux-reversal DC-field machine for wind power generation. IEEE Trans. Energy Convers. **30** (4), 1565–1573 (2015)
7. D.U. Campos-Delgado, D.R. Espinoza-Trejo, E. Palacios, Fault-tolerant control in variable speed drives: a survey. IET Electr. Power Appl. **2**(2), 121–134 (2008)
8. M. Cheng, W. Hua, J. Zhang, W. Zhao, Overview of stator-permanent magnet brushless machines. IEEE Trans. Industr. Electron. **58**(11), 5087–5101 (2011)
9. Y. Pang, Z.Q. Zhu, D. Howe, Analytical determination of optimal split ratio for permanent magnet brushless motor. IET Electr. Power Appl. **153**(1), 7–13 (2006)
10. Y. Wang, M. Cheng, M. Chen, Y. Du, K.T. Chau, Design of high-torque-density double-stator permanent magnet brushless motors. IET Electr. Power Appl. **5**(3), 317–323 (2011)
11. C. Lee, R. Krishnan, N.S. Lobo, Novel two-phase switched reluctance machine using common-pole E-core structure: concept, analysis, and experimental verification. IEEE Trans. Ind. Appl. **45**(2), 703–711 (2009)
12. P. Zheng, Q. Zhao, J. Bai, B. Yu, Z. Song, J. Shang, Analysis and design of a transverse-flux dual rotor machine for power-split hybrid electric vehicle applications. Energies **6**(12), 6548–6568 (2013)
13. Y. Wang, K.T. Chau, C.C. Chan, J.Z. Zhang, Transient analysis of a new outer-rotor permanent-magnet brushless DC drive using circuit-field-torque time-stepping finite element method. IEEE Trans. Magn. **38**(2), 1297–1300 (2012)
14. Y. Fan, K.T. Chau, M. Cheng, A new three-phase doubly salient permanent magnet machine for wind power generation. IEEE Trans. Ind. Appl. **42**(1), 53–60 (2006)

Part III
A Comprehensive Study of Advanced Magnetless Machines on Electric and Hybrid Vehicles

Chapter 8
Proposed Axial-Field Machine for Range-Extended Electric Vehicles

8.1 Introduction

As suggested in the previous chapters, the electric vehicle (EV) applications have attracted many attentions to cope with the energy issues [1]. In general, the traditional EVs employ only the battery as the pure energy source so this type of EVs can only support for short duration trips and impractical practical for standard drivers [2]. To extend the driving range for standard applications, the range-extended EV (RE-EV) that implements an internal combustion engine (ICE) as secondary energy source has become very attractive [3, 4].

Even though the permanent-magnet (PM) brushless machines have been widely employed in the RE-EV system, the advanced magnetless doubly salient (DS) machines with DC-field excitation have been gaining more attention recently [5]. In the meantime, with the adaption of the radial length as the active part to generate torque, the axial-field (AF) machine can significantly improve its torque density, as compared with its profound radial-field (RF) counterparts [6, 7]. However, the incorporation of the AF structure and the magnetless DC-field machine has not been suggested yet.

The purpose of this chapter is to newly implement the DC-field excitation into the AF structure, and thus forming a novel axial-field doubly salient DC-field (AF-DSDC) machine, purposely for RE-EV propulsion system. The design criteria and principle of operation for the proposed AF-DSDC machine will be covered. The machine performances will be analyzed with the help of the 3D finite element method (3D-FEM), and then quantitatively compared with the requirements of standard RE-EV systems. To achieve a better presentation, the commonly employed RF machines, namely the RF-switched reluctance (RF-SR) machine, the RF-DSDC machine and the RF-doubly salient permanent-magnet (RF-DSPM) machine are also included for extensive comparisons.

© Springer Nature Singapore Pte Ltd. 2018
C. H. T. Lee, *Design, Analysis and Application of Magnetless Doubly Salient Machines*, Springer Theses, https://doi.org/10.1007/978-981-10-7077-8_8

8.2   System Architecture and Principle of Operation

The typical RE-EV architecture, which applies the in-wheel direct-drive topology, is shown in Fig. 8.1. Based on the requirements of a standard RE-EV system, the targeted specifications of the proposed system for each motors are tabulated in Table 8.1 [8]. The electric machine is employed as the major energy source for the RE-EV system as normal operation, while the ICE is instead utilized as the implemented component as range extension operation. In general, four major operating principles can be provided by the proposed RE-EV system.

8.2.1   Torque Boosting

In the occasion if the RE-EV has to speed up or climb uphill, the machine needs to undergo the torque boosting mode, namely the machine has to generate the maximum torque within a short duration. Based on this operation mode, the DC-field current density is purposely increased to strengthen the flux densities, so that the peak output torque is generated.

Fig. 8.1 Range-extended electric vehicle architecture

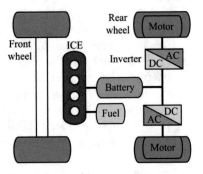

Table 8.1 Targeted motor specifications for the RE-EV system

Item	Value
Peak DC voltage (V)	360
Rated power (kW)	4.7
Rated torque (N m)	150
Constant-torque operation (rpm)	0–300
Constant-power operation (rpm)	300–900
Peak torque	130% for 10 s
Wheel dimension	195/65 R15

8.2.2 Flux-Weakening Operation

When the RE-EV has already started up based on the torque boosting mode, the generated electromotive force (EMF) of the corresponding armature windings start to increase in accordance to the higher operating speed. To retain a definite power level over the constant power region, the flux density has to be weakened accordingly. Based on the regulation of the controllable DC-field current, the proposed RE-EV system can efficiently accomplish the flux-weakening characteristics. The RE-EVs can theoretically achieve infinite operating ranges with control of the DC-field currents.

8.2.3 Fault-Tolerant Operation

If the DC-field excitation experiences the open-circuit fault or short-circuits fault conditions, the DC-field winding has to be cut off such that the proposed machine can still provide the basic function based on the remaining healthy armature winding sets [9]. When the DC-field current is not working properly, the remaining armature current can be reconfigured accordingly and operated based on the unipolar conduction algorithm, and hence the normal torque value can be retained. In the meantime, with the adaption of the traditional fault-tolerant principles, the proposed machine can still achieve the fault-tolerant operation for the armature winding fault. The fault tolerability is extremely critical for RE-EV system because it can prevent traffic jams or severe accidents if the propelling machine suddenly experiences fault conditions.

8.2.4 Battery Charging

Once the energy of battery is almost used up, the ICE can be employed to operate the RE-EV and to recharge up the battery in order to extend the operating range. Because the operating speed of ICE changes in accordance to vehicle speeds, the output voltage is changing as well. In the meantime, if the RE-EV goes downhill or during braking situation, the electric machine experiences a regenerative braking scenario to recharge the battery [10]. This time-changing output voltage can overcharge the battery and deteriorate its cycle life. To lengthen the battery life, the proposed machine can regulate its DC-field current to control the flux density such that the output voltage can be kept as constant in a wide operating range.

8.3 Proposed Axial-Field Doubly Salient DC-Field Machine

8.3.1 Proposed Machine Structure

The structure of the proposed 8/10-pole AF-DSDC machine that contains a double-sided-stator sandwiched-rotor topology is shown in Fig. 8.2. The construction of the proposed structure is similar to those employed on the other traditional AF counterparts [6]. With the proposed structure and the proper dimensions, the rotors of the proposed AF machine can be mount on the tire seamlessly, and hence the in-wheel motor direct-drive structure can be instinctively accomplished as shown in Fig. 8.3.

The proposed machine is designed with the four-phase topology because of the considerations between the stability and cost-effectiveness. The key machine design data are categorized in Table 8.2. Unlike the RF counterparts, the proposed AF-DSDC machine utilizes its radial part for torque production and hence its torque density can be improved significantly.

Because the AF-DSDC machine is derived from the traditional RF-DSDC machine, its design equations can be derived from that of the profound RF-DSDC machine. Hence, the pole arrangements of the AF-DSDC machine can be described as follows

$$\begin{cases} N_s = 2mk \\ N_r = N_s \pm 2k \end{cases} \tag{8.1}$$

where N_s is the number of stator poles, N_r the number of rotor poles, m the number of armature phases and k any integer. By selecting $m = 4$ and $k = 1$, this ends up with $N_s = 8$ and $N_r = 10$, and hence the proposed structure for the AF-DSDC machine is resulted.

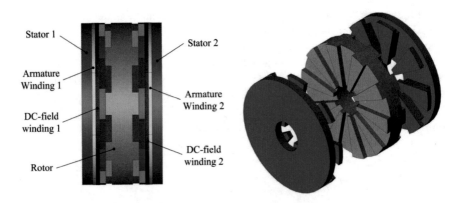

Fig. 8.2 Proposed axial-field doubly salient DC-field machine

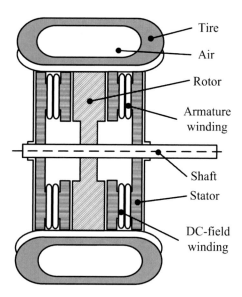

Fig. 8.3 In-wheel direct-drive internal topology

Table 8.2 Key data of the proposed AF-DSDC machine

Item	AF-DSDC
Radial outside diameter (mm)	381
Radial inside diameter (mm)	100
Axial stack length (mm)	195
Air-gap length of both segments (mm)	0.5
Number of stator poles of both segments	8
Number of rotor poles of both segments	10
Number of armature phases	4
Number of turns per armature coil	50
Rotor and stator material	Steel sheet: 50JN700 (JFE Steel Corporation, Tokyo, Japan)
Armature and DC-field winding material	Copper

8.3.2   DC Flux-Linkage Pattern

The proposed AF-DSDC machine installs two winding types, namely the DC-field winding and armature winding, on its sided-stators, while two winding sets both employ the concentrated winding configuration. Based on this arrangement, each of

Fig. 8.4 DC flux-linkage
paths of the proposed
AF-DSDC machine.
a Position 1. **b** Position 2

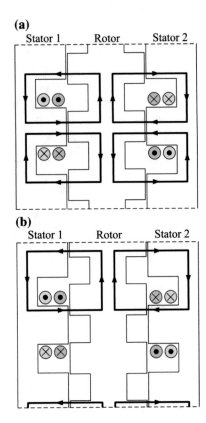

them is allocated in a way that the corresponding magnetic axes are parallel to each
other, with opposing directions. These arrangement results the divided fluxes as two
equal paths in the rotor yoke as shown in Fig. 8.4.

Since the AF-DSDC machine consists of a symmetrical structure, the power
transferred by the two sided-stators can be perfectly integrated with each other, and
hence the balanced resultant torque is accomplished. To ease the control complexity, each of the two corresponding sided-stators winding sets can be purposely
connected in series.

8.3.3 Operating Principles

Based on the controllable DC-field winding, the proposed AF-DSDC machine can
be operated with two different conduction algorithm, namely the bipolar conduction
algorithm and unipolar conduction algorithm. In particular, the bipolar conduction
algorithm mainly serves as the normal operations, while the unipolar conduction
algorithm as the fault-tolerant operation. A full-bridge inverter is employed to

control the armature current, so that the independent phase control functionality is achieved. In the meantime, a full-bridge DC-DC converter is instead employed for the DC-field current, and hence the flux strengthening and weakening capabilities are provided.

8.3.3.1   Bipolar Conduction Algorithm

In the scenario if the DC-field current works properly, the proposed AF-DSDC machine can employ the bipolar conduction algorithm, which is similar to the algorithm employed in the RF-DSPM machine, as normal operation [11]. During the increasing period of the DC flux-linkage Ψ_{DC} where the no-load EMF is positive, a positive current I_{BLDC} is injected to the armature winding to generate a positive torque $T_{DC}.$ On the other hand, a negative current $-I_{BLDC}$ is instead injected during the decreasing period of the Ψ_{DC} where the no-load EMF is negative, and hence the positive torque is generated as well. The theoretical waveforms based on the bipolar conduction algorithm are shown in Fig. 8.5a.

Each armature phase experiences the 90° conduction where $\theta_2 - \theta_1 = \theta_4 - \theta_3 = 90°$. The generated electromagnetic torque T_{DC} can be described as

Fig. 8.5 Principle of operations. **a** Bipolar conduction algorithm. **b** Unipolar conduction algorithm

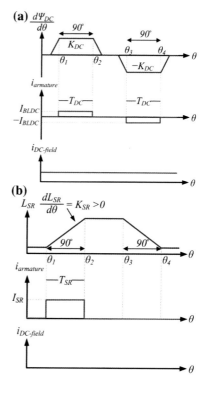

$$T_{DC} = \frac{1}{2\pi} \int_0^{2\pi} \left(i_{BLDC} \frac{d\psi_{DC}}{d\theta} + \frac{1}{2} i_{BLDC}^2 \frac{dL_D}{d\theta} \right) d\theta \qquad (8.2)$$

where L_D is the self-inductance. Based on this conduction algorithm, the generated torque component is mainly contributed by the DC-field excitation, while the torque component generated by the reluctance principle is relatively minor with a zero average value [11]. Therefore, the pulsating zero reluctance torque can be neglected and the torque equation can be further expanded as

$$T_{DC} = \frac{1}{2\pi} \left(\int_{\theta_1}^{\theta_2} I_{BLDC} K_{DC} d\theta + \int_{\theta_3}^{\theta_4} (-I_{BLDC})(-K_{DC}) d\theta \right) = \frac{1}{2} I_{BLDC} K_{DC} \qquad (8.3)$$

where K_{DC} is the slope of Ψ_{DC} with respect to θ.

8.3.3.2 Unipolar Conduction Algorithm

In the case when the DC-field excitation experiences any fault situation, the DC-field winding should be terminated such that the proposed machine can employ the unipolar conduction algorithm, which is also employed by the traditional RF-SR machine. To be specific, a unipolar armature rectangular current I_{SR} is injected when the self-inductance L_{SR} is increasing such that the positive reluctance torque T_{SR} is generated during $\theta_2 - \theta_1 = 90°$, as shown in Fig. 8.5b. Yet, based on this conduction algorithm, only half of the useful zone is adapted for torque production. Hence, the torque performance is degraded and the torque pulsation is relatively more severe than that generated based on the bipolar conduction algorithm. Thus, this conduction algorithm can be utilized as the fault-tolerant mode for the situation if the DC-field excitation experiences the fault conditions. The reluctance torque generated based on the unipolar conduction algorithm can be described as

$$T_{SR} = \frac{1}{2\pi} \int_0^{2\pi} \left(\frac{1}{2} i_{SR}^2 \frac{dL_{SR}}{d\theta} \right) d\theta = \frac{1}{2\pi} \int_{\theta_1}^{\theta_2} \left(\frac{1}{2} I_{SR}^2 K_{SR} \right) d\theta = \frac{1}{8} I_{SR}^2 K_{SR} \qquad (8.4)$$

where K_{SR} is the slope of L_{SR} with respect to θ. To retain the same torque level between two principles, the Eqs. (8.3) and (8.4) should be equated to yield

$$I_{SR} = 2 \sqrt{\frac{I_{BLDC} K_{DC}}{K_{SR}}} \qquad (8.5)$$

According to Eq. (8.5), the armature current at the unipolar conduction algorithm can be derived, and hence the proposed machine can generate the same torque level as compared with the torque at the bipolar conduction algorithm. Yet, the unipolar conduction algorithm typically needs the armature current with higher magnitude. Hence, the unipolar operation is not desirable for normal operations. Furthermore, the unipolar conduction algorithm adapts only half of the useful zone for torque production. Thus, the corresponding torque pulsation is larger than the bipolar conduction counterpart.

8.4   Electromagnetic Field Analysis

Electromagnetic field analysis is one of the most famous development tools for electric machines, and it can be generally classified into two major type, namely the analytical field type [12] and the numerical field type [13]. In this chapter, 3D-FEM is adapted for the performance analysis for the proposed AF-DSDC machine. JMAG-Designer is employed to perform the finite element analysis for the proposed machine, while more than 10 hours are needed to finish one simulation process based on a high-performance PC. The no-load magnetic flux distributions of the proposed machine are shown in Fig. 8.6. The magnetic flux is divided into two symmetrical patterns and flow through the rotor part to the two sided-stators. Therefore, the two individual torques generated by the two segments are expected to be equal in nature.

The airgap flux density of the two airgaps of the proposed AF-DSDC machine with the no-load condition under 5 A/mm^2 DC-field current as the rated condition is shown in Fig. 8.7. The results confirm the two airgaps can accomplish the equal pattern. Therefore, based on the equal electric supply, the generated torques by the two sided-stators are expected to be equal.

The DC flux-linkage waveforms based on the bipolar conduction algorithm (with 5 A/mm^2 DC-field current) and the self-inductance waveforms based on the unipolar conduction algorithm (with zero DC-field current) are shown in Figs. 8.8 and 8.9, respectively.

To retain the equal torque value between two conduction algorithms, the armature current for the unipolar algorithm can be deduced by Eq. (8.5) where the values of K_{DC} and K_{SR} can be figured out based on Figs. 8.8 and 8.9, as tabulated in Table 8.3. To be specific, the armature currents at the unipolar conduction algorithm need to be increased by about 129%, as compared with the bipolar conduction counterpart. With this particular current level, the proposed machine can still function for a relatively long duration, say around a couple of hours without severe damages.

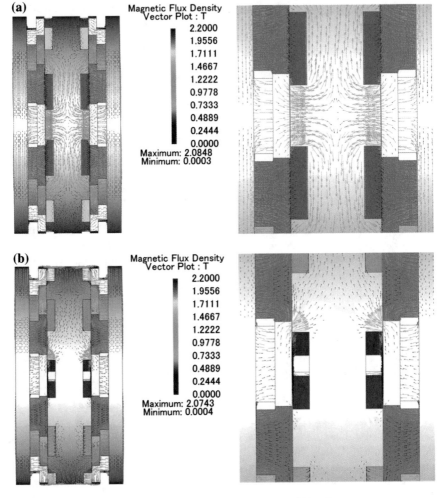

Fig. 8.6 No-load magnetic flux distributions. **a** Position 1. **b** Position 2

8.5 Performances Analysis

8.5.1 No-Load EMF Waveforms at Rated Conditions

With the employment of 3D-FEM, the machine performances of the proposed AF-DSDC machine can be deduced. At the beginning, the no-load EMF waveforms under the base speed of 300 rpm based on the 5 A/mm² rated DC-field current are calculated and as shown in Fig. 8.10. The no-load EMF waveforms of two winding sets both provide the four-phase characteristics with balanced patterns. The results verify the pole-pair configuration and the winding settings are correct. Furthermore,

Fig. 8.7 Airgap flux
densities. **a** Airgap 1.
b Airgap 2

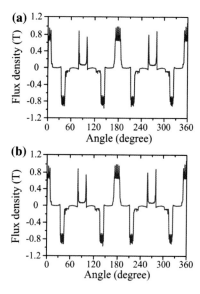

Fig. 8.8 DC flux-linkage
waveforms. **a** Winding 1.
b Winding 2

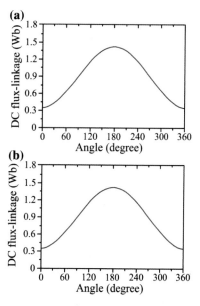

it can be seen that the no-load EMF waveforms offer equal polarities as well as
magnitudes, with two of them can approach to the maximum magnitude of 153 V.
Therefore, the targeted DC supply voltage is still capable to sustain the normal
function even if the two armature winding sets are connected in series.

Fig. 8.9 Self-inductance
waveforms. **a** Winding 1.
b Winding 2

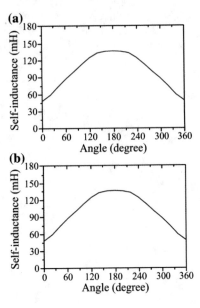

Table 8.3 Key parameters
for two conduction algorithms
of AF machine

Item	Winding 1	Winding 2
Rated current at bipolar conduction (A)	10	10
Slope of DC flux-linkage K_{DC}	1.04	1.04
Slope of self-inductance K_{SR}	0.079	0.079
Rated current at unipolar conduction (A)	22.9	22.9

8.5.2 Performance of Mode 1—Torque Boosting

The torque performances of the proposed AF-DSDC machine at 5 A/mm^2 rated
DC-field current and at 10 A/mm^2 strengthened DC-field current, are shown in
Fig. 8.11. Because the controllable DC-field current can be regulated easily, the
airgap flux densities can be varied effectively to fulfil various scenarios. The
average steady torques of the proposed machine at rated and strengthened DC-field
currents are 154.2 and 203.2 N m, respectively. The results verify that the proposed
machine can comply with the torque standard for the typical driving conditions,
including the *Mode 1*, the torque boosting mode. In addition, the torque pulsating
values at rated and strengthened DC-field currents can be calculated as 27.4 and
43.2%, respectively, where two of them are within the acceptable range for standard
RE-EVs [3].

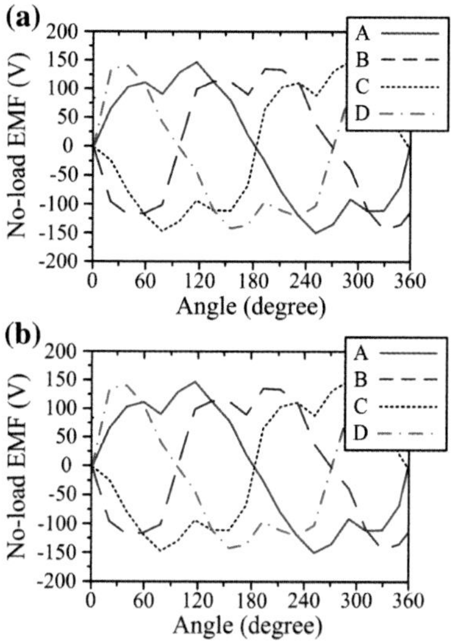

Fig. 8.10 No-load
electromotive force
waveforms at rated
conditions. **a** Winding 1.
b Winding 2

Fig. 8.11 Torque waveforms
at bipolar conduction
algorithm. **a** Rated DC-field
current. **b** Strengthened
DC-field current

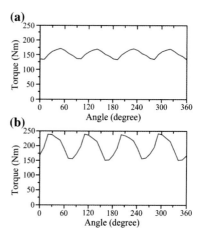

In addition, the cogging torque waveforms of the proposed machine at rated and strengthened DC-field currents are calculated and as shown in Fig. 8.12. The maximum magnitudes at rated and strengthened DC-field currents are found to be 18.6 and 42.4 N m, respectively. These values are only 12.1 and 20.9%, as compared with the corresponding average torques, respectively. In particular, the results confirm that the cogging torque values are also acceptable as compared with the commonly employed RF-DSPM counterparts.

Fig. 8.12 Cogging torque
waveforms at bipolar
conduction algorithm. **a** Rated
DC-field current.
b Strengthened DC-field
current

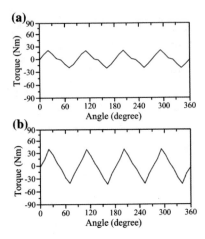

8.5.3   Performance of Mode 2—Flux-Weakening Operation

The torque-speed characteristics of the proposed machine are calculated and shown
in Fig. 8.13. The DC-field current can be separately controlled so that the proposed
machine can offer effective flux-weakening function to achieve the wide operating
range, as one of the key considerations for RE-EVs. As confirmed, the proposed
machine can retain the constant power characteristics over the spread of speed
range. To be specific, the machine speed can reach to about 907 rpm when the
DC-field current is tuned to be 1.5 A/mm^2. This result has confirmed that the
proposed machine can offer the targeted operating range and comply with the
standard for the *Mode 2*, the flux-weakening operation.

Fig. 8.13 Torque-speed
characteristics of the proposed
machine

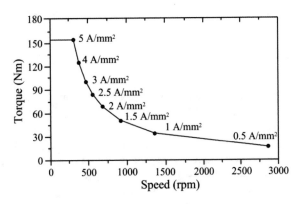

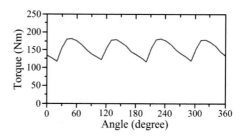

Fig. 8.14 Torque waveforms with the unipolar conduction algorithm

8.5.4 Performance of Mode 3—Fault-Tolerant Operation

In the case if the DC-field current is terminated, the proposed machine can still provide the basic function based on the unipolar conduction algorithm and the corresponding steady torque is shown in Fig. 8.14. The averaged torque and the torque pulsation with the unipolar conduction algorithm are about 148.2 N m and 40.1%, respectively. The results verify that when the DC-field current is cut off, the proposed machine can still provide practically equal torque value for basic function with the unipolar conduction algorithm. In the meantime, since only half of the useful period is employed, the torque pulsations as well as the armature current values are larger as compared with the torque generated by the bipolar conduction algorithm. In addition, the efficiency at the unipolar conduction algorithm is relatively worse than that happened at the bipolar conduction counterpart. Therefore, the unipolar conduction algorithm should be employed as fault-tolerant mode or so-called as the *Mode 3*.

8.5.5 Performance of Mode 4—Battery Charging

Finally, Fig. 8.15 shows the no-load EMF waveforms based on different operating speeds with two scenarios, namely without the flux control and with the flux control, respectively. As expected, the values of the generated voltages change in accordance to the varying speeds, when there is no DC-field regulation. This varying output voltage is not desirable for battery charging. In the meantime, the generated voltages can be maintained as constant value over a wide operating range, when there is DC-field regulation. Hence, this particular characteristic can then comply with the requirements for the *Mode 4*, the battery charging mode.

8.6 Comparisons with Radial-Field Machines

The performances of the proposed AF-DSDC machine are calculated and deduced based on the 3D-FEM and tabulated in Table 8.4.

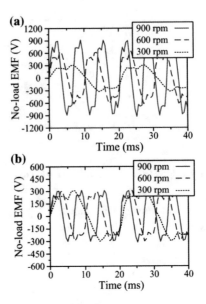

Fig. 8.15 No-load EMF waveforms at various speeds. **a** Without flux control. **b** With flux control

Table 8.4 Proposed AF-DSDC machine performances

Item	Bipolar conduction	Unipolar conduction
Rated power (kW)	4.8	4.6
Power density (W/kg)	35.4	33.8
Operating frequency (Hz)	50	50
No-load EMF of Winding 1 (V)	153	N/A
No-load EMF of Winding 2 (V)	153	N/A
Rated DC-field excitation (A/mm^2)	5	N/A
Rated torque (N m)	154.2	148.2
Torque ripple at rated torque (%)	27.4	40.1
Cogging torque at rated torque (N m)	18.6	N/A
% cogging torque at rated torque (%)	12.1	N/A
Boosted DC-field excitation (A/mm^2)	10	N/A
Boosted torque (N m)	203.2	N/A
Torque ripple at boosted torque (%)	43.2	N/A
Cogging torque at boosted torque (N m)	42.4	N/A
% cogging torque at boosted torque (%)	20.9	N/A

To further emphasize the advantages of the proposed AF-DSDC machine, the common RF machines, namely the RF-SR machine, the RF-DSDC machine and the RF-DSPM machine are included for comparisons, as shown in Fig. 8.16a, b, c, respectively. To provide a fair environment, all suggested machines share the equal

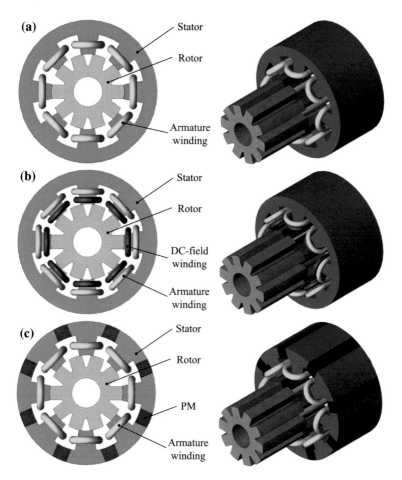

Fig. 8.16   Radial-field machines. **a** RF-SR machine. **b** RF-DSDC machine. **c** RF-DSPM machine

Table 8.5   Comparisons between axial-field and RF machines

Item	RF-SR	RF-DSDC	RF-DSPM	AF-DSDC
Rated torque/mass (N m/kg)	0.51	0.61	1.21	1.13
Rated torque/volume (kN m/m³)	4.12	4.85	9.51	8.97
Material cost (USD)	208.4	209.8	411.9	239.5
Rated torque/cost (N m/USD)	0.34	0.41	0.39	0.65

key dimensions, namely the radial outside diameters, radial inside diameters, axial stack lengths and airgap lengths. Furthermore, the slot area factors, the winding factors as well as the current densities are also set to be equal.

With the help of the FEM, the key performances of the RF machines are simulated and tabulated in Table 8.5. Not surprisingly, the magnetless RF-SR machine utilizes only half of its useful zone for torque production such that the poorest performances, including the worst gravimetric and volumetric torque densities, are resulted. In the meantime, the RF-DSDC machine, which employs all the useful zones, can provide the better performances as compared with the RF-SR machine. Upon the employment of the radial part for torque production, the proposed AF-DSDC machine can provide greater torque densities as compared with the common magnetless RF counterparts, and is particularly comparable to the RF-DSPM machine.

Despite the RF-DSPM machine can generate the greatest torque densities than the magnetless counterparts, the supply of the PM materials is pulsating such that the PM materials cost has increased significantly [14]. To improve the market penetration of the RE-EV applications, the cost-effectiveness should also be included as the design criteria. Referring to the latest market prices of the raw materials, the materials costs of the machines can be calculated. Based on the PM-free structure, the proposed AF-DSDC machine accomplishes the best performance with respect to the cost-effectiveness among all the counterparts, and hence this proposed machine is expected to be desirable for the RE-EV applications.

8.7   Summary

In this chapter, a new magnetless AF-DSDC machine is proposed, purposely for RE-EV system. With the employment of the radial active part to produce torque, the proposed AF-DSDC machine provides better performance than the traditional RF counterparts do, in terms of the cost-effectiveness. Based on the regulation on the controllable DC-field current, the proposed machine can achieve various operating principles, namely the torque boosting, the flux-weakening, the fault-tolerant and the battery charging modes. Therefore, the proposed machine is expected to provide good perspective for the RE-EV applications.

References

1. M. Ehsani, K.M. Rahman, H.A. Toliyat, Propulsion system design of electric and hybrid vehicles. IEEE Trans. Industr. Electron. **44**(16), 19–27 (1997)
2. C.C. Chan, The state of the art of electric and hybrid vehicles. Proc. IEEE **90**(2), 247–275 (2002)
3. M. Ehsani, Y. Gao, J.M. Miller, Hybrid electric vehicles: architecture and motor drives. Proc. IEEE **95**(4), 719–728 (2007)
4. K.T. Chau, C.C. Chan, Emerging energy-efficient technologies for hybrid electric vehicles. Proc. IEEE **95**(4), 821–835 (2007)
5. K.T. Chau, M. Cheng, C.C. Chan, Nonlinear magnetic circuit analysis for a novel stator doubly fed doubly salient machine. IEEE Trans. Magn. **38**(5), 2382–2384 (2002)
6. F. Profumo, Z. Zhang, A. Tenconi, Axial flux machines drives: a new viable solution for electric cars. IEEE Trans. Industr. Electron. **44**(1), 39–45 (1997)

7. C.H.T. Lee, K.T. Chau, C. Liu, T.W. Ching, F. Li, A high-torque magnetless axial-flux doubly-salient machine for in-wheel direct drive applications. IEEE Trans. Magn. **50**(11), 8202405 (2014)

8. M. Olszewski, *Evaluation of the 2010 Toyota Prius Hybrid Synergy Drive Systems* (Oak Ridge National Laboratory, U.S. Department Energy, Washington, DC, USA, 2011)

9. C. Yu, K.T. Chau, New fault-tolerant flux-mnemonic doubly-salient permanent-magnet motor drive. IET Electr. Power Appl. **5**(5), 393–403 (2009)

10. C. Liu, K.T. Chau, J.Z. Jiang, A permanent-magnet hybrid brushless integrated-starter-generator for hybrid electric vehicles. IEEE Trans. Industr. Electron. **57**(12), 4055–4064 (2010)

11. J. Zhang, M. Cheng, Z. Chen, W. Hua, Comparison of stator-mounted permanent-magnet machines based on a general power equation. IEEE Trans. Energy Convers. **24**(4), 826–834 (2009)

12. P. Zheng, Q. Zhao, J. Bai, B. Yu, Z. Song, J. Shang, Analysis and design of a transverse-flux dual rotor machine for power-split hybrid electric vehicle applications. Energies **6**(12), 6548–6568 (2013)

13. Y. Wang, K.T. Chau, C.C. Chan, J.Z. Zhang, Transient analysis of a new outer-rotor permanent-magnet brushless DC drive using circuit-field-torque time-stepping finite element method. IEEE Trans. Magn. **38**(2), 1297–1300 (2012)

14. M. Chen, K.T. Chau, W. Li, C. Liu, Cost-effectiveness comparison of coaxial magnetic gears with different magnet materials. IEEE Trans. Magn. **50**(2), 7020304 (2014)

Chapter 9
Proposed Reliable Gearless Machine for Magnetic Differential System

9.1 Introduction

Electric vehicles (EVs) have been regarded as the cleanest and greenest road transportation for smart cities [1, 2]. Traditionally, based on a mechanical differential (MD) system, the two wheels of EVs are driven by a single traction machine to provide both straight motion and curvilinear trajectory movement [3, 4]. However, this approach employs a heavy and bulky differential gear and it is absolutely unfavorable for the modern EV system [5]. To get rid of this heavy, bulky and inefficient MD system, two or even more machines are separately installed as the driving wheels. As a result, each of the wheels can be independently controlled to achieve different speeds during turning, and this is known as the electronic or electric differential (ED) system [6, 7]. With the support of the ED system, the differential gear can be removed such that the overall weight and transmission loss of EVs can be drastically reduced. Nevertheless, even the ED system can provide the differential action by operating individual machines independently [8], fatal accidents might occur when there are any control or feedback errors from the separated machines.

To maintain the advantages of high safety and robustness of the MD system and attain the merits of high compactness and accuracy of the ED system simultaneously, the concept of magnetic steering machines was suggested [9]. However, the corresponding operating principles and the analysis of desired differential action are still absent in literature.

In recent years, the development of magnetless doubly salient machines have been speeding up, while the flux-switching DC (FSDC) topology is the most popular candidate [10, 11]. Since the limited supply of rare-earth permanent-magnets (PMs) and the high robustness of rotor structure, this machine type is becoming very attractive to serve as in-wheel gearless machines for EV applications.

© Springer Nature Singapore Pte Ltd. 2018
C. H. T. Lee, *Design, Analysis and Application of Magnetless Doubly Salient Machines*, Springer Theses,
https://doi.org/10.1007/978-981-10-7077-8_9

The purpose of this chapter is to develop a new class of reliable gearless machines that adopts magnetic steering (MS) to realize the concept of the magnetic differential (MagD) system for direct-drive EVs. With the implementation between the MS winding and the double-rotor (DR) FSDC machine, a new MS-DR-FSDC machine is proposed. The MS-field winding is utilized to generate an additional field excitation in such a way that it reacts to the command of steering wheel. Consequently, it produces the MS flux to magnetically interlock the magnetic fields between the two spinning rotors. During cornering, with the presence of MS flux, the resultant flux at one rotor is strengthened while at another rotor is weakened. As a result, the two rotors can be driven with different torques to achieve the differential action for turning. Because the MS flux offers the interlock mechanism with the magnetic mean, the proposed MagD system can absolutely provide higher reliability than the conventional ED system.

9.2 Proposed Magnetic Differential System

9.2.1 Vehicle Dynamic Modeling

Over the years, the dynamic modeling of engine vehicles has been well developed [12], while the corresponding model can be extended to EV system seamlessly. In particular, the total tractive force F_{tract} of the EV is to provide an acceleration force F_a, while overcoming the total resistive force F_{res}. The corresponding modeling can be described as

$$F_{tract} = F_a + F_{res} \tag{9.1}$$

$$\begin{cases} F_{res} = F_{roll} + F_{aero} + F_{slope} \\ F_{roll} = \mu Mg \\ F_{aero} = \dfrac{1}{2} \rho C_x S v^2 \\ F_{slope} = Mg \sin \alpha \end{cases} \tag{9.2}$$

where F_{roll} is the rolling resistance, F_{aero} is the aerodynamic drag force, F_{slope} is the slope resistance, μ is the rolling friction coefficient, M is the total mass of the EV, ρ is the air density, C_x is the aerodynamic drag coefficient, S is the frontal area of the EV, v is the linear speed of the EV and α is the slope of the road.

9.2.2   *Curvilinear Movement*

When the EV has to turn an angle, the exterior wheels have to run with a larger radius than the interior wheels do such that the ED system has to drive the wheels at two different speeds. To be specific, the exterior wheels have to spin at higher speeds than those of the interior ones. To analyze the curve manoeuver mathematically, the simplified geometry model [7] is employed as shown in Fig. 9.1. For instance, when the EV has to turn right, the linear speeds of two individual wheels v_L and v_R can be described as

$$\begin{cases} v_L = \omega_{curve}\left(r + \dfrac{d}{2}\right) \\ v_R = \omega_{curve}\left(r - \dfrac{d}{2}\right) \end{cases} \tag{9.3}$$

where $\omega_{curve} = v/r$ is the angular speed of the curve, v is the linear speed of the EV, r is the radius of the curve and d is the width of the EV. The radius of curve r can be further expressed as

$$r = \frac{l}{\tan \delta} \tag{9.4}$$

where l is the length of the EV and δ is the turning angle.

When the EV is about to turn an angle, the steering signal is applied in a way the exterior wheel has to increase its speed while the interior wheel has to decrease it accordingly. The difference between their angular speeds $\Delta\omega$ can be described as

Fig. 9.1 Geometry model of an electric vehicle

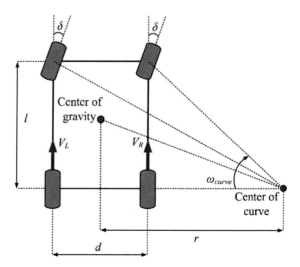

$$\Delta\omega = \frac{d \tan \delta}{l} \omega_{vehicle} \tag{9.5}$$

where $\omega_{vehicle}$ is the angular speed of the EV. Consequently, the two speed references for the left and right wheels can be expressed as

$$\begin{cases} \omega_L* = \omega_L + \dfrac{\Delta\omega}{2} \\ \omega_R* = \omega_R - \dfrac{\Delta\omega}{2} \end{cases} \tag{9.6}$$

For the ED system as shown in Fig. 9.2a, the curvilinear trajectory movement can be achieved with the independent control of two wheels. The two speed references ω_L* and ω_R* are fed into two speed controllers separately, so that two independent torques T_L and T_R are generated to drive the left and right wheels, respectively. Even though the ED system can provide the differential action to turn an angle, it suffers from the bulky structure when two machines are installed for propulsion and the degraded reliability when independent control of two wheels are applied.

For the proposed MagD system as shown in Fig. 9.2b, the curvilinear trajectory movement can be achieved instead with only one speed controller and one machine. In particular, the speed difference reference between the two wheels $\Delta\omega*$ is fed into the speed controller, so that the MS-field current signal I_f* is generated to control

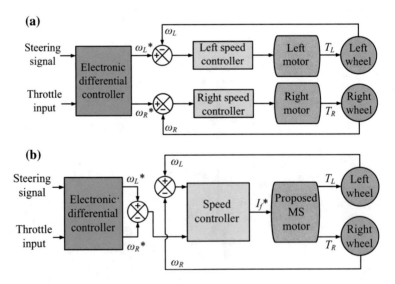

Fig. 9.2 Differential systems. **a** ED. **b** Proposed MagD

the MS-field excitation. Upon the regulation of the MS-field excitation, the proposed MS machine can therefore differentiate its two output torques T_L and T_R, so as the wheel speeds, with the magnetic means. As compared with the ED counterpart, the proposed system can greatly improve the machine compactness and system reliability.

9.3   Proposed Magnetic Steering Machines

9.3.1   Proposed Machine Structures

To thoroughly assess the proposed MS concept, two basic morphologies are studies —namely the radial-field (RF) and the axial-field (AF) morphologies. The structures of the RF-MS-DR-FSDC machine and the AF-MS-DR-FSDC machine are shown in Figs. 9.3 and 9.4, respectively. Both of them adopt the same pole arrangement, which are extended from the profound three-phase 12/10-pole FSDC

Fig. 9.3 Proposed RF-MS-DR-FSDC machine. **a** Configuration. **b** 3D view

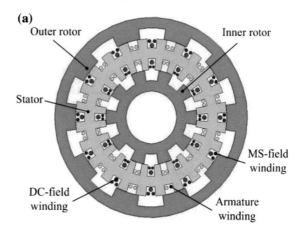

(a)

Outer rotor

Inner rotor

Stator

MS-field winding

DC-field winding

Armature winding

(b)

Fig. 9.4 Proposed
AF-MS-DR-FSDC machine.
a Configuration. **b** 3D view

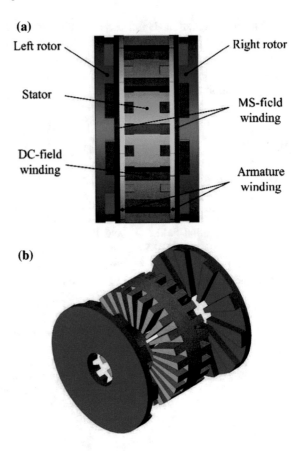

(a)

Left rotor — Right rotor

Stator —

MS-field
winding

DC-field
winding

Armature
winding

(b)

machine. The major distinction between the two machines comes from the flux flow directions. In particular, the flux path of the RF-MS-DR-FSDC machine flows along the radial direction, while that of the AF-MS-DR-FSDC machine along the axial direction. Upon the profound difference in structure, the RF-MS-DR-FSDC machine is developed with the sandwiched-stator concentric-rotor structure, while the AF-MS-DR-FSDC machine with the sandwiched-stator sided-rotor structure.

Similar to other FSDC machines, the two proposed machines consist of the armature winding and DC field winding in the stator, while their rotors consist of simple iron core with barely salient poles. Unlike other FSDC machines, the proposed machines incorporate also the distinctive MS-field winding for magnetic interlocking. The MS-field winding is purposely installed in such a way that it can offer flux strengthening and flux weakening operations with respect to the two armature windings in the stator. Consequently, the two rotors can be driven with different torques, and hence the differential torque when the MS-field excitation is applied. As a result, the differential action can be achieved based on one control parameter.

Since the proposed two machines are purposely developed for in-wheel machine drive applications, both of them are designed based on a typical wheel size of EVs. For instance, the wheel size of 195/65 R15, i.e., the axial length is 195 mm and the rim diameter is 15 in. (381 mm), is employed. To have a fair comparison, the key machine dimensions, namely the outside diameters, inside diameters, stack lengths, airgap lengths and slot-fill factors of the two proposed machines are set equal. In the meantime, the pole arcs, pole heights and current densities are optimized in a way to minimize magnetic saturation and the core losses.

9.3.2   Operating Principles

The proposed RF-MS-DR-FSDC and the AF-MS-DR-FSDC machines employ the concentrated winding arrangements on the sandwiched-stator in such a way that the excited flux by the DC-field windings, indicated as the DC flux, flow along the two rotors via the sandwiched-stator, as shown in Fig. 9.5. In the meantime, the flux excited by the MS-field winding, indicated as the MS flux, can superimpose with the DC flux, in order to create the flux-strengthening effect at one side while the flux-weakening effect at another side. This is the distinctive mechanism to make use

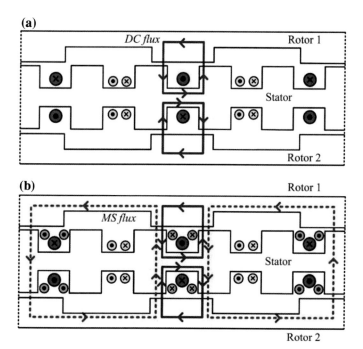

Fig. 9.5 Excitation fluxes in MS machines. **a** Without MS-field for straight motion. **b** With MS-field for cornering

of the MS flux to generate the differential torque for the differential action during cornering. In principle, the DC-field windings can be wound with the concentrated winding arrangement to directly regulate the magnetic fields of two rotors individually. Nevertheless, without the excitation of MS-field winding, the two rotors will operate independently and without interconnection. As a result, the magnetic interlock mechanism will not exist, and it cannot offer the higher reliability. In addition, in inheriting the flux-switching characteristic, the flux-linkages of these machines reverse their polarities with the rotor positions accordingly. Therefore, the iron core material can be fully utilized to produce higher power and torque densities.

Since the inner and outer peripheral areas of the stator of the RF-MS-DR-FSDC machine are different, their electromagnetic interaction with the inner and outer rotors are different in nature. On the other hand, the left and right peripheral areas of the stator of the AF-MS-DR-FSDC machine are identical, so that their electromagnetic interactions with the left and right rotors should be the same. Consequently, the AF-MS-DR-FSDC machine undoubtedly provides more desirable balancing properties between the two rotors than the RF-MS-DR-FSDC machine.

9.3.3   Torque Production

Depending the airgap flux waveforms, the proposed MS machines can be operated at either the brushless AC (BLAC) mode or the brushless DC (BLDC) mode [4]. Because the BLAC operation can offer smoother torque than the BLDC operation, the proposed machines have been optimized to offer more sinusoidal flux-linkage Ψ. As shown in Fig. 9.6, when the DC-field winding current $I_{DC\text{-}field}$ is applied, the proposed machines can generate a positive electromagnetic torque T at the scenario when the armature current $I_{armature}$ is injected with respected to Ψ. In the meantime,

Fig. 9.6 Theoretical operating waveforms without MS-field

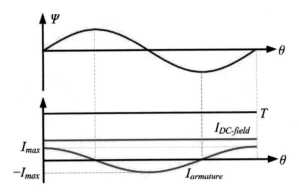

because of nature of doubly salient poles, the reluctance torque components are resulted accordingly. Since these reluctance torque components are relatively small with an averaged zero value, it can be assumed that the generated torques are contributed only by the DC-field components.

When the EV needs to curve an angle, the MS-field winding has to be excited accordingly. The excited MS flux can then superimpose with the DC flux, creating flux strengthening at one side and flux weakening at another side. The polarity of the MS-field excitation can determine which side to be strengthened or weakened. Consequently, the torque production mechanism at each rotor is the same, while the magnitude of torque difference between two rotors relies only on the magnitude of the MS-field excitation. Because the airgap flux densities at two rotors are regulated in a conjugated manner, a magnetically interlocking mechanism between the two rotors can be accomplished. As a result, higher reliability can be achieved in the proposed MagD system.

9.4 Machine Performance Analysis

9.4.1 Magnetic Field Distributions and Flux-Linkages

The key design data of the proposed MS machines are listed in Table 9.1. With the support of a commercial finite element method (FEM) software package, the JMAG-Designer, the electromagnetic field analysis can be performed to evaluate the major performances of the proposed machines.

The airgap flux densities of the AF-MS-DR-FSDC machine are shown in Fig. 9.7. It should be noted that three situations are considered, namely without MS-field excitation, with positive MS-field and with negative MS-field excitation.

Table 9.1 Key design data of proposed MS machines

Items	MS-RF-FSDC	MS-AF-FSDC
Radial outside diameter (mm)	381.0	381.0
Radial inside diameter (mm)	100.0	100.0
Airgap length (mm)	0.5	0.5
Axial stack length (mm)	195.0	195.0
No. of stator poles	12	12
No. of rotor poles	10	10
No. of phases	3	3
Slot-fill factor	60%	60%
No. of outer armature turns	52	N/A
No. of inner armature turns	38	N/A
No. of left armature turns	N/A	46
No. of right armature turns	N/A	46

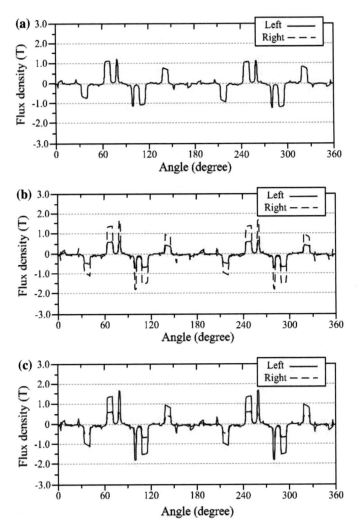

Fig. 9.7 Airgap flux densities of AF-MS-DR-FSDC machine. **a** Without MS-field. **b** With positive MS-field. **c** With negative MS-field

Since the two machines produce similar flux patterns, to improve the readability, the airgap flux densities of the RF-MS-DR-FSDC machine are purposely neglected. It can be shown that if MS-field excitation is absent, the AF-MS-DR-FSDC machine can produce identical airgap flux densities between its two sides. On the other hand, if the MS-field winding is positively excited, the flux density of the right airgap is strengthened, while that of the left airgap is instead weakened. The situation is reversed if the MS-field winding is negatively excited, i.e., the right side is weakened and the left side is strengthened. Similar results are given at the RF-MS-DR-FSDC machine, where the inner and outer airgap fluxes act as the left

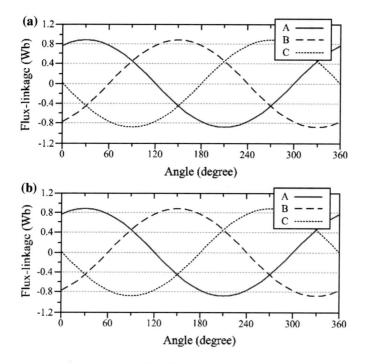

Fig. 9.8 Flux-linkage waveforms of AF-MS-DR-FSDC. **a** Left. **b** Right

and right airgap fluxes. Hence, the FEM results well agree with the theoretical expectation as shown in Fig. 9.5. As a result, the validity of the proposed MS-field regulation mechanism is confirmed.

Next, the flux-linkages of the proposed AF-MS-DR-FSDC machine with only the DC-field excitation of 5 A/mm^2 are simulated as shown in Fig. 9.8. It can be shown that the flux-linkages in each winding set are well balanced among three phases with bipolar pattern. These results show the proposed machines can assemble with that of the conventional FSDC machine does. Consequently, the major design criteria such as the pole arrangement and the winding allocation are confirmed as correct.

It should be mentioned the values of flux-linkage between the two armature windings in the AF-MS-DR-FSDC machine are identical. It is because the peripheral areas of the left and right sides of the stator are identical in structure. On the other hand, it can be predicted the values of flux-linkage between the two armature windings in the RF-MS-DR-FSDC machine should be different because its corresponding peripheral areas are different.

9.4.2 No-Load Electromotive Forces (EMF) Performance

The no-load electromotive forces (EMFs) of the two proposed machines with only the DC-field excitation of 5 A/mm^2 at base speed of 300 rpm can be predicted easily from the flux-linkage waveforms. As expected, the two proposed machines can both produce no-load EMF with symmetrical patterns. In the meantime, the no-load EMF waveforms should be very sinusoidal in nature, and it gives evidence to show these machines are favorable for BLAC operation.

The RF-MS-DR-FSDC machine consists of different peripheral area between its inner stator side and outer stator side, namely the peripheral area of the inner side is smaller than that of the outer side. Consequently, inheriting from the different flux-linkage values between the inner and outer sides, the RF-MS-DR-FSDC machine produces the no-load EMF with different values between two sides. On the other hand, since the peripheral areas of the left and right sides are the same, the no-load EMF magnitudes of the AF-MS-DR-FSDC machine are identical.

9.4.3 Output and Cogging Torques

The output torque waveforms of the proposed machines with the DC-field excitations of 5 A/mm^2 are simulated as shown in Fig. 9.9. It can be shown that the rated average torques at the outer and inner rotors of the RF-MS-DR-FSDC machine are 110.2 and 59.8 N m, respectively. The corresponding torque ripples are 18.5 and 15.2%, respectively. In the meantime, the rated average torques at the left and right rotors of the AF-MS-DR-FSDC machine are 132.5 and 132.2 N m, respectively. The corresponding torque ripples are 21.5 and 21.4%, respectively. The results verify the two generated torques of the AF-MS-DR-FSDC machine are exactly the same, and this characteristic is very suitable for the proposed MagD system.

In addition, to provide a more comprehensive analysis on the torque performances, the cogging torque waveforms are also provided in Fig. 9.10. It can be observed the peak magnitude of cogging torque in the RF-MS-DR-FSDC machine are 7.6 and 3.8 N m, respectively, while those of the AF-MS-DR-FSDC machine are 10.9 and 10.7 N m, respectively. These magnitudes equal 6.9 and 6.4% of its rated torque magnitudes for the RF-MS-DR-FSDC machine and 8.2 and 8.1% for the AF-MS-DR-FSDC machine. The cogging torques of the two proposed machines are relatively high because the designs have not undergone any process for cogging torque reduction yet. Basically, these cogging torque values can be minimized based on the rotor skewing structure [13] or pole-arc optimization [14].

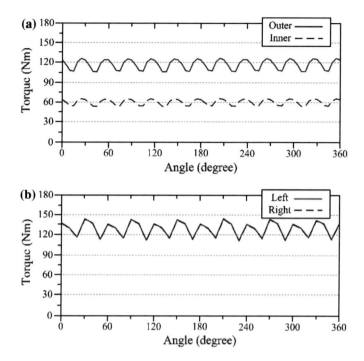

Fig. 9.9 Rated torque waveforms of proposed machines without MS-field. **a** RS-MS-DR-FSDC. **b** AF-MS-DR-FSDC

9.4.4 Differential Torques

Based on the aforementioned analysis, for the proposed MagD system, it is obviously that the AF-MS-DR-FSDC machine is more desirable than the RF-MS-DR-FSDC machine because of the torque balancing characteristic between two rotors. Consequently, the emphasis is given to the AF-MS-DR-FSDC machine on the performance evaluations for the proposed MagD system.

The output torque waveforms of the AF-MS-DR-FSDC machine when the MS-field windings are excited are shown in Fig. 9.11. It can be observed that when the MS-field winding is excited positively of 5 A/mm^2, the average torques of the left and right rotors result with 107.1 and 142.7 N m, respectively, and thus corning a left angle. In contrast, when the MS-field winding is excited negatively of 5 A/mm^2, the average torques of the left and right rotors result with 142.5 and 106.8 N m, respectively, and thus corning a right angle. Hence, the results verifty that the proposed MS-AF-FSDC machine can generate the same torque difference, or known as the differential torque, between the two rotors for various curvilinear movement.

Because of magnetic saturation, when it compares to the torque values at the situation without MS-field excitation as shown in Fig. 9.9b, the torque increments

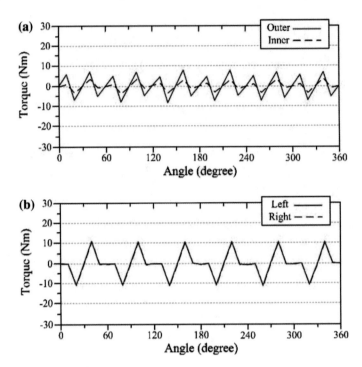

Fig. 9.10 Cogging torque waveforms of proposed machines without MS-field. **a** RS-MS-DR-FSDC. **b** AF-MS-DR-FSDC

under the flux-strengthening effect are slightly less than the torque decrements under the flux-weakening effect. However, this will not cause any problem to the proposed MagD system because the turning capability depends only on the differential torque between the two rotors.

Furthermore, the differential torque versus MS-field excitation characteristic of the proposed AF-MS-DR-FSDC machine is shown in Fig. 9.12. It can be observed that the relationship is fairly linear and this characteristic is highly favorable for accurate steering control. To be specific, with the 5 A/mm^2 MS-field excitation, the differential torque can reach 35.6 N m and it is very effective for normal turning conditions.

9.5 System Performance Analysis

9.5.1 Curvilinear Motion Under Normal Operation

According to Eqs. (9.1–9.6), the curvilinear motion of an EV can be calculated with the support of MatLab/Simulink. The major parameters of the EV model are listed

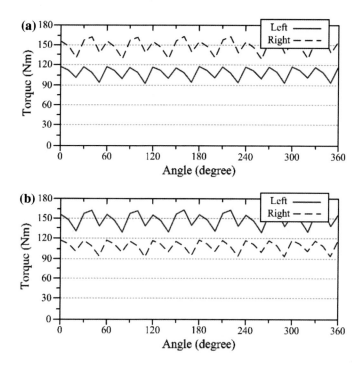

Fig. 9.11 Output torque waveforms of AF-MS-DR-FSDC machine with MS-field. **a** Positive excitation. **b** Negative excitation

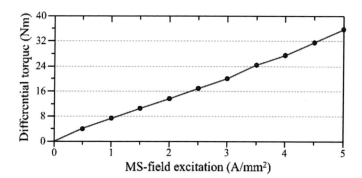

Fig. 9.12 Differential torque versus MS-field excitation characteristic of AF-MS-DR-FSDC machine

in Table 9.2. When the machines are under normal conditions, the vehicular performance comparisons between the ED system and the proposed MagD system with various turning angles are shown in Fig. 9.13.

Table 9.2 Key parameters of EV model

Parameters	Value
Total mass, M (kg)	1200
Rolling friction coefficient, μ	0.015
Air density, ρ (kg/m^3)	1.184
Aerodynamic drag coefficient, Cx	0.25
Frontal area, S (m^2)	1.9
Vehicle width, d (m)	1.5
Vehicle length, l (m)	2.5
Slope angle, α (°)	0

For the ED system, the curvilinear motion can be accomplished with the independent control of the individual torques of two separated machines based on different turning angles. In particular, at $\delta = 60°$, the EV is turning right with an angle of 60°, while $\delta = -60°$ means the EV is instead turning left with an angle of 60°. As shown, the speeds of the two wheels are differentiated accordingly while the EV is maintained at the constant speed of 30 km/h.

As compared with the ED system, the proposed MagD system can achieve the same result based on the control of the MS-field current I_f. To be specific, I_f can be utilized as the single parameter to differentiate the two torques of the two wheels. As illustrated in Fig. 9.13d, when $I_f < 0$, the driving torque at the left rotor is greater than that at the right rotor, such that the EV is turning a right angle. On the other hand, when $I_f > 0$, the EV is then turning a left angle. Finally, when $I_f = 0$, the EV will keep as a straight line motion. Generally speaking, different turning angles can be accomplished with different values of I_f.

9.5.2 Curvilinear Motion Under Fault Condition

Even though the proposed MagD system can achieve the same differential action as the ED system does, it is essential to justify its distinctive advantage regarding the reliability. To offer a fair comparison, both of the ED system and the proposed MagD system are subject to the same machine fault at the same situation, i.e., the EV is turning a right angle of 60° at the time instant of 8 s. For the ED system, this machine fault occurs at either one of the two machines (for example, the left machine); while for the proposed MagD system, the MS machine is subjected to the same fault conditions. It should be noted that various types, such as the short-circuit fault, open-circuit fault or inter-turn fault, can happen in the fault conditions.

The vehicular performance comparisons between the ED system and the proposed MagD system under the same machine fault are shown in Fig. 9.14. To be specific, the machine can no longer provide adequate torque level to fulfill the load torque during fault conditions, so that the corresponding speed decreases consequently.

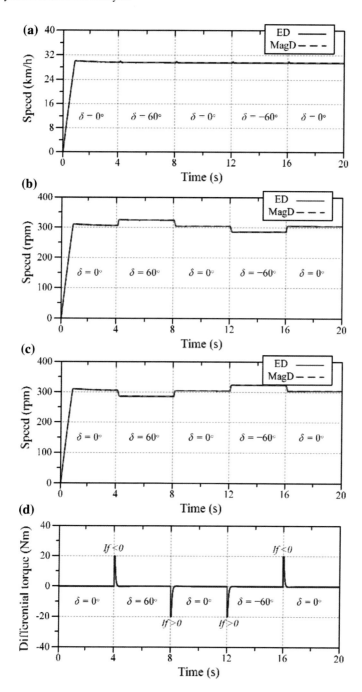

Fig. 9.13 Vehicular performances of ED system versus MagD system under normal operation. **a** Speed of EV. **b** Speed of left wheel. **c** Speed of right wheel. **d** Differential torque of MagD system

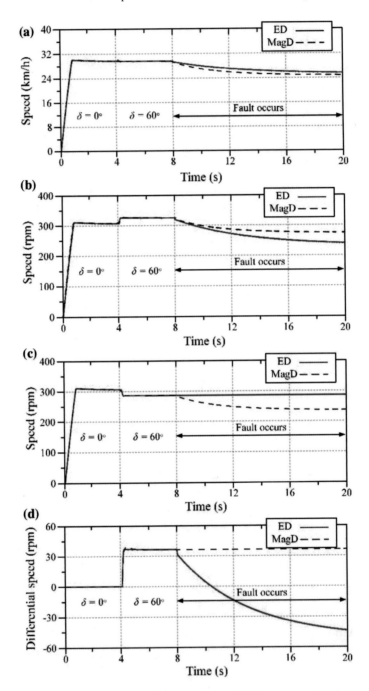

Fig. 9.14 Vehicular performances of ED system versus MagD system under fault condition.
a Speed of EV. **b** Speed of left wheel. **c** Speed of right wheel. **d** Differential speed of two wheels

For the ED system, the speed of left wheel decreases while the speed of right wheel keeps unchanged, so that the speed of EV start decreasing at the time instant of 8 s. Consequently, the differential speed between the left wheel and right wheel decreases from positive to negative as shown in Fig. 9.14d. As a result, the EV changes from turning a right angle to a left angle, and it may cause a fatal accident.

On the other hand, for the proposed MagD system, both the speed of left wheel and right wheels decrease concurrently because both rotors are magnetically interlocked by the turning angle. As illustrated, the differential speed of the two wheels are maintained at the definite value. Consequently, the EV is still turning a right angle correctly even if the EV speed has decreased when there is a machine fault. As a result, it confirms that the proposed MagD system is much more reliable than the conventional ED system.

9.6 Summary

In this chapter, a new class of reliable gearless machines, known as the MS machine, has been proposed and employed to realize the MagD system for direct-drive EV applications. The distinct mechanism is to utilize the magnetic interlocking technique of the proposed MS machine to allow the MagD system to achieve the differential action for turning. To be specific, two proposed MS machines, namely the RF-MS-DR-FSDC machine and the AF-MS-DR-FSDC machine, are quantitatively analyzed. Based on the regulations of MS-field excitation, the two rotors can be differentiated to produce appropriate torques to drive two separate wheels. Between the two MS machines, the AF-MS-DR-FSDC machine enjoys the merits of better torque density and higher balancing, so that it is more desirable for the MagD system. Through the system level simulation, it verifies the proposed MagD system can achieve similar curvilinear motion performance as the ED system under normal operation. Most importantly, the proposed MagD system can retain the differential action under fault conditions, which might cause a fatal accident when the ED system is instead employed.

References

1. A. Emadi, Y.J. Lee, K. Rajashekara, Power electronics and machine drives in electric, hybrid electric, and plug-in hybrid electric vehicles. IEEE Trans. Ind. Electron. **55**(6), 2237–2245 (2008)
2. W. Hua, G. Zhang, M. Cheng, Investigation and design of a high-power flux-switching permanent magnet machine for hybrid electric vehicles. IEEE Trans. Magn. **51**(3), 8201805 (2015)
3. G. Tao, Z. Ma, L. Zhou, L. Li, A novel driving and control system for direct-wheel-driven electric vehicle. IEEE Trans. Magn. **41**(1), 497–500 (2005)

4. K.T. Chau, *Electric Vehicle Machines and Drives-Design, Analysis and Application* (Wiley, New York, NY, USA, 2015)
5. J. Kim, C. Park, S. Hwang, Y. Hori, H. Kim, Control algorithm for an independent machine-drive vehicle. IEEE Trans. Veh. Technol. **59**(7), 3213–3222 (2010)
6. N. Mutoh, T. Kazama, K. Takita, Driving characteristics of an electric vehicle system with independently driven front and rear wheels. IEEE Trans. Ind. Electron. **53**(3), 803–813 (2006)
7. A. Draou, Electronic differential speed control for two in-wheels machine drive vehicle, in *Proceedings Fourth International Conference Power Engineering, Energy and Electrical Drives*, pp. 764–769, May 2013
8. R. Wang, J. Wang, Fault-tolerant control with active fault diagnosis for four-wheel independently driven electric ground vehicles. IEEE Trans. Veh. Technol. **60**(9), 4276–4287 (2011)
9. C.H.T. Lee, K.T. Chau, C. Liu, A new magnetic steering axial-field machine for electronic differential system in electric vehicle, in *Proceedings International Magnetic Conference 2015*, Paper No. AV–12, May 2015
10. Y. Tang, J.J.H. Paulides, T.E. Motoasca, E.A. Lomonova, Flux-switching machine with dc excitation. IEEE Trans. Magn. **48**(11), 3583–3586 (2012)
11. C.H.T. Lee, K.T. Chau, C. Liu, A high-torque magnetless axial-flux doubly-salient machine for in-wheel direct drive applications. IEEE Trans. Magn. **50**(11), 8202405 (2014)
12. R.N. Jazar, *Vehicle Dynamics: Theory and Applications* (Springer, New York, NY, USA, 2008)
13. W. Fei, P.C.K. Luk, J. Shen, Torque analysis of permanent-magnet flux switching machines with rotor step skewing. IEEE Trans. Magn. **48**(10), 2664–2673 (2012)
14. C.H.T. Lee, K.T. Chau, C. Liu, Design and analysis of an electronic-geared magnetless machine for electric vehicles. IEEE Trans. Ind. Electron. **63**(11), 6705–6714 (2016)

Chapter 10
Proposed Electronic-Geared Machine for Electric Vehicle Applications

10.1 Introduction

Energy utilization and environmental protection have become the hot research topics in recent years, while as one of the most promising solutions to improve the current situations, the development of the electric vehicles (EVs) is speeding up [1, 2]. As the key component of the EV technologies, electric machines have to offer high efficiency, high power density, high controllability, wide-speed range, maintenance-free operation and fault-tolerant capability [3, 4]. The doubly salient permanent-magnet (DSPM) machine with unipolar-flux characteristic in particular can achieve most of the tasks, and hence it has drawn many attentions in the past few decades [5]. Meanwhile, inheriting the bipolar-flux characteristic and thus resulting higher power density, the flux-switching permanent-magnet (FSPM) machine has become very popular [6, 7].

The PM machines undoubtedly offer a great potential for EV propulsion, yet the PM candidates suffer from the problems of high PM material costs and uncontrollable PM flux densities [8]. To overcome the profound shortcomings of PM machines, the advanced magnetless machines, which are cost-effective and flux-controllable, have become popular recently [9]. Moreover, in order to cater different extreme operating situations, the concept of dual-mode operations [10] for magnetless machines has been described in Chap. 7. Hence, the dual-mode machine can operate at the low-speed and high-speed operations to capture the maximum power for weak and strong winds, respectively. However, the corresponding machine exhibits unbalanced flux-linkages, leading to higher torque ripple, which is intolerable for EV propulsion system.

This chapter aims to propose a new magnetless machine, known as the electronic-geared (EG) machine, purposely for EV propulsion. Following the concept of dual-mode principle, the proposed machine is developed based on the multi-tooth bipolar-flux (MTBF) operation and the single-tooth unipolar-flux (STUF) operation. To be specific, the former operation mode can be employed for

© Springer Nature Singapore Pte Ltd. 2018
C. H. T. Lee, *Design, Analysis and Application of Magnetless Doubly Salient Machines*, Springer Theses,
https://doi.org/10.1007/978-981-10-7077-8_10

the low-gear (high-torque low-speed) situation, while the latter one for the high-gear (low-torque high-speed) situation. The balance-position winding arrangement, which can achieve balanced flux-linkages, will be newly implemented into the proposed machine, so that the desired torque performance can be achieved. Moreover, the machine will be designed in such a way that the back electromotive force (EMF) waveforms can facilitate both the MTBF and STUF operations. The machine performances will be analyzed thoroughly by using the finite element method (FEM), with emphasis on validating the proposed electronic gearing concept. In addition, the experimental setup is developed for machine design verification.

10.2   Proposed Electronic-Geared Magnetless Machine

10.2.1   Machine Structure

Figure 10.1 shows the machine structure of the proposed EG magnetless machine for EV propulsion. It artfully combines the design criteria of the MTBF machine [11] and STUF machine [5]. Hence, it can inherit the corresponding machine characteristics and operate with two operations, namely the MTBF and the STUF operations. The design criteria of both the MTBF and the STUF machines are governed by the general equations as follows

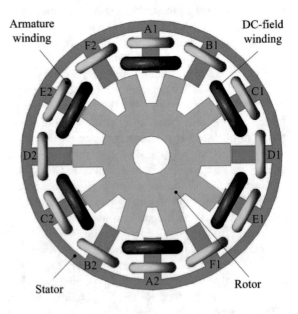

Fig. 10.1 Electronic-geared magnetless machine structure

Armature winding

DC-field winding

Stator

Rotor

$$\begin{cases} N_{sp} = 2mk \\ N_{se} = N_{sp}N_{st} \\ N_r = N_{se} \pm 2k \end{cases} \tag{10.1}$$

where N_{sp} is the number of stator poles, N_{st} the number of stator teeth, N_{se} the number of equivalent stator poles, N_r the number of rotor poles, m the number of armature phases and k any integer.

Even though the MTBF and STUF machines share the same design criteria of Eq. (10.1), they can be distinguished by the numbers of armature phases and stator teeth. In particular, with the same numbers of equivalent stator poles and rotor poles, the bipolar-flux and the unipolar-flux machines can be actualized by the multi-tooth and the single-tooth arrangements, respectively. Namely, the multi-tooth least-phase machine is suitable for low-speed operation, while the single-tooth multi-phase machine is instead favorable for high-speed operation.

In order to simultaneously realize two types of machines, the number of equivalent stator poles and the number of rotor poles, i.e., N_{se} and N_r, of the two machines should be equalized. However, the derived relationship results with infinite number of solutions. Hence, to reduce the degree of freedom and produce a unique solution, the value of k between the two machines is purposely equalized and the relationship is further derived as

$$m'N_{st}' = m''N_{st}'' \quad (N_{st}' > N_{st}'') \tag{10.2}$$

where m' and N_{st}' are the numbers of armature phases and stator teeth for the MTBF machine, respectively, while m'' and N_{st}'' are the numbers of armature phases and the stator teeth for the STUF machine, respectively. Based on the Eqs. (10.1) and (10.2), the fundamental design combinations of the proposed EG machine can be obtained as listed in Table 10.1.

To ease the control complexity and to minimize the cost of power electronic devices, the least numbers of armature phases are purposely chosen, i.e., three-phase for MTBF operation and six-phase for STUF operation. Moreover, to simplify the manufacturing process, the least numbers of stator and rotor poles are preferred. By taking these criteria into consideration, the combination of $k = 1$, $m' = 3$, $N_{st}' = 2$, $m'' = 6$, $N_{st}'' = 1$, $N_{se} = 12$ and $N_r = 10$ is selected as the structure of the proposed EG machine. Based on this combination, the proposed machine is anticipated to have the larger torque value at the MTBF operation, whereas the extended operating ranges at the STUF operation.

Table 10.1 Fundamental design combinations of the EG machine

k	m'	N_{st}'	m''	N_{st}''	N_{se}	N_r	N_r
1	3	2	6	1	12	14	10
1	3	4	6	2	24	26	22
1	4	2	8	1	16	18	14
2	3	2	6	1	24	28	20

10.2.2 Winding Arrangement

The proposed EG magnetless machine adopts the concentrated winding arrangement, which allows it to have higher connection flexibility, i.e., each of the armature winding coil can be controlled independently, and hence different connection arrangements can be configured to fulfill the criteria of different operating modes.

In Chap. 7, the multi-tooth operation was achieved by connecting its multiple phases in the adjacent positions. With this adjacent-position winding arrangement, the flux-linkages among armature phases are unbalanced, and hence the back EMF waveforms end up with the asymmetry patterns, which are unfavorable for the torque production. In particular, these asymmetric back EMF waveforms will produce larger torque ripple, and hence resulting in undesirable acoustic noise and vibration problems.

To improve the situation, the proposed EG magnetless machine at the MTBF operation is purposely connected with the so-called the balance-position winding arrangement, i.e., A1, A2, D1 and D2 are connected in series; B1, B2, E1 and E2 in series; C1, C2, F1, and F2 in series. Based on the proposed winding arrangement, the flux-linkages among the armature phases are balanced, and hence the back EMF waveforms can become more symmetric than its counterpart does.

10.2.3 Back EMF Waveforms

For the EV propulsion systems, some important issues should be considered. Namely, the torque ripple should be minimized at the low-speed high-torque operation, while the torque density should be increased at the high-speed low-torque operation. To take these criteria into considerations, the back EMF waveforms should be carefully analyzed.

In general, the back EMF waveform of electric machines can be classified into two major types, namely the sinusoidal-like waveform and the trapezoidal-like waveform [12]. To effectively operate these machines, there are two bipolar conduction schemes available, namely the brushless AC (BLAC) and the brushless DC (BLDC) schemes, for the sinusoidal-like and the trapezoidal-like machines, respectively.

For the sinusoidal-like machine, the sinusoidal armature current I_{BLAC} is applied in accordance to the status of the flux-linkage Ψ_{BLAC} to produce the positive electromagnetic torque T_{BLAC}. This BLAC conduction scheme is depicted in Fig. 10.2a. With the BLAC conduction scheme, the sinusoidal-like machine can perfectly match the injected armature currents with its back EMF waveforms, and hence achieving the minimized torque ripple performance.

For the trapezoidal-like machine, a positive rectangular current I_{BLDC} is applied to the armature winding when the flux-linkage Ψ_{BLDC} is increasing so as to produce

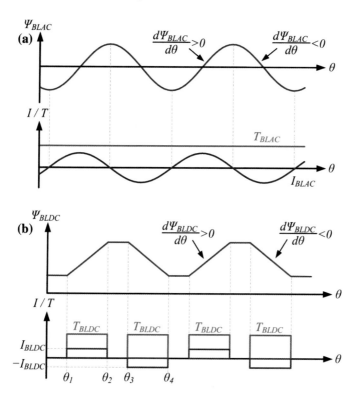

Fig. 10.2 Theoretical operating principles. **a** BLAC conduction scheme. **b** BLDC conduction scheme

the positive torque T_{BLDC}. Meanwhile, a negative current is instead applied to the armature winding when the flux-linkage is decreasing so as to produce also the positive torque. This BLDC conduction scheme is depicted in Fig. 10.2b. With the BLDC conduction scheme, the trapezoidal-like machine can offer higher torque density as compared with its sinusoidal-like counterpart does.

10.2.4 Operating Principles

According to the aforementioned discussions, the proposed EG machine should be designed in a way to offer the no-load EMF waveforms showing in between the sinusoidal-like and the trapezoidal-like characteristics. Based on this design criterion, the proposed machine at the MTBF operation can be operated similarly as the sinusoidal-like machine does, and thus producing the torque with minimized pulsation in the low-speed situation. Meanwhile, the proposed machine at the STUF operation can be instead operated similarly as the trapezoidal-like machine does, and hence providing the improved torque density in the high-speed situation.

At the MTBF operation, the proposed machine behaves similarly as the three-phase sinusoidal-like machine does and it can be operated by the BLAC operation scheme with the armature currents of the three-phase sinusoidal form as

$$
\begin{cases}
i_a = I_{MTBF} \sin\theta \\
i_b = I_{MTBF} \sin(\theta + (2\pi/3)) \\
i_c = I_{MTBF} \sin(\theta - (2\pi/3))
\end{cases}
\tag{10.3}
$$

where $i_{a,b,c}$ and I_{MTBF} are the corresponding armature currents and the maximum value of the phase currents, respectively, at the MTBF operation.

At the STUF operation, the proposed machine behaves similarly as the six-phase trapezoidal-like machine does and it can be operated by the BLDC operation scheme. To maintain the same input power level as that employed in the MTBF operation, the magnitudes of the phase current at the STUF operation should be reduced accordingly and given as

$$
\begin{cases}
i_k = I_{STUF} & \theta_1 \le \theta \le \theta_2 \\
i_k = 0 & 0 < \theta < \theta_1, \theta_2 < \theta < \theta_3, \theta_4 < \theta < 2\pi \\
i_k = -I_{STUF} & \theta_3 \le \theta \le \theta_4
\end{cases}
\tag{10.4}
$$

where i_k and I_{STUF} are the corresponding armature currents and the maximum value of phase currents, respectively, at STUF operation. According to the proposed operating principles, the proposed machine can offer relatively smoother torque at MTBF operation, while higher torque density at STUF operation.

10.2.5   Analysis of the Operating Range Extension

Unlike the PM machines, the magnetless machines can utilize its controllable DC-field excitation for flux regulation, and hence offering the flux-weakening characteristics, to extend its operating ranges effectively. However, the flux-weakening capability can never cover the infinite operating ranges and there are some constraints behind.

Undoubtedly, the flux-linkage varies along with the flux regulation, i.e., when the DC-field excitation is weakened, the flux-linkage decreases accordingly. However, in the meantime, the self-inductance also gradually increases [5]. Based on this phenomenon, at a particular point, the DC-field electromagnetic torque no longer serves as the major torque component, while the reluctance torque will replace its dominating position. In other word, the machine has to operate with the reluctance principle, where only half of the torque producing zone is utilized.

Upon this scenario, larger torque ripple is resulted and larger armature current is needed to maintain the same torque level, which are both unfavorable for EV applications.

Compared with the DC-field operation, the reluctance operating principle should serve as the fault-tolerant operation. To avoid the machine to switch to the reluctance operation, the DC-field excitation should be kept at certain levels. However, it will reduce the flux-weakening capability and limit the operating ranges. To relieve the undesirable dilemma, the proposed machine can instead incorporate the mode-changing principle with the flux-weakening operation, for better range extension performances.

10.2.6 Proposed Control Scheme

The control scheme for the proposed dual-mode EG machine is shown in Fig. 10.3, and it can be divided into five major parts, namely (i) the armature controller, (ii) the armature inverter, (iii) the DC-field controller, (iv) the DC-field converter and (v) the EG machine.

The armature controller adopts the profound dual-closed-loop control scheme, i.e., the outer speed loop employs a PID regulator for speed control, and the inner current loop adopts a hysteresis regulator for current chopping control. Based on the comparison on the speed command $n*$ and the actual speed n, the armature current command $i*$ is generated. Meanwhile, based on the comparison between $i*$ and the actual current i, the hysteresis regulator signal can be further resulted.

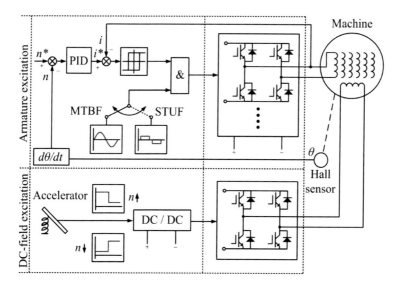

Fig. 10.3 Control scheme for EG operation

Consequently, the firing signal of each power switch in the armature inverter can be developed. In addition to the conventional dual-closed-loop control, the armature controller requires a more complicated inverter for the armature winding that favors the complex control circuit, i.e., the inverter can be switched to provide the BLAC current for the MTBF operation, while the BLDC current for the STUF operation.

To offer the flux regulating capability, the DC-field excitation module that consists of two major components, namely the DC/DC converter and the H-bridge inverter, is adopted. In particular, the DC/DC converter is used to regulate the excitation level, while the H-bridge converter to control the direction of the current.

10.3   Machine Performance Analysis

10.3.1   Electromagnetic Field Analysis

To study the performances of the electric machines, the electromagnetic field analysis has been recognized as one of the most accurate and convenient tools for many years. In this chapter, the JMAG-Designer is employed as the magnetic solver to perform the FEM. The magnetic field distribution of the proposed machine at no-load condition is shown in Fig. 10.4, while the result shows that the flux distributions are well balanced and align with the theoretical results.

The airgap flux density waveform of the proposed machine at the no-load condition is shown in Fig. 10.5. The result shows that the airgap flux density is within the normal range, and hence providing additional evidence to illustrate that

Fig. 10.4 Magnetic field distribution of the EG machine

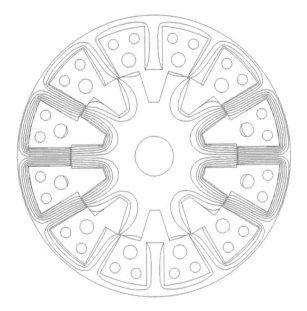

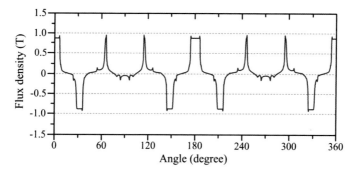

Fig. 10.5 Airgap flux density of the EG machine

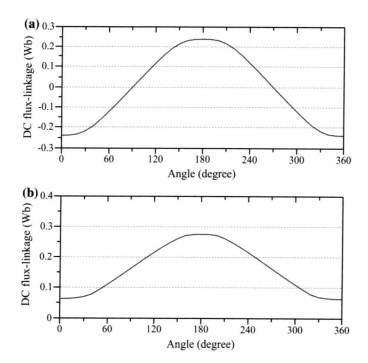

Fig. 10.6 Flux-linkage waveforms of EG machine. **a** MTBF operation. **b** STUF operation

the proposed machine design can surpass the magnetic saturation, in order to minimize the core losses.

The flux-linkage waveforms at the MTBF and at the STUF operations are shown in Fig. 10.6a, b, respectively. The simulation waveforms show the proposed machine can offer the bipolar flux-linkage at MTBF operation, while the unipolar flux-linkage at STUF operation. The results confirm that the armature winding arrangement and the phase selection of the proposed machine are both correct.

10.3.2   Pole-Arc Ratio Analysis

The characteristics of the no-load EMF waveforms can be modified by the so-called pole-arc ratio p, while p is defined as the ratio of the rotor pole-arc β_r to the stator pole-arc β_s, i.e., $p = \beta_r/\beta_s$. In order to minimize the magnetic saturation and to maximize the armature slot areas, the β_s is first set as a particular value. Then, the β_r is selected as equal to the value of β_s, i.e., $p = 1.0$, as shown in Fig. 10.7a, in the beginning stage. After that, the value of p can be modified by tuning the value of β_r, and hence achieving the optimal pole-arc ratio as: $p\_opt = \beta_{r\_opt}/\beta_s$, as shown in Fig. 10.7b.

The variations of the no-load EMF waveforms according to different values of p are shown in Fig. 10.8. As discussed, the proposed machine should be designed to offer the no-load EMF waveforms obtaining the characteristics in between the sinusoidal-like and the trapezoidal-like patterns. Hence, the pole-arc ratio should be selected between $p = 1.2$ and 1.3.

Fig. 10.7 Pole-arc ratio variations. **a** Primitive case. **b** Optimal case

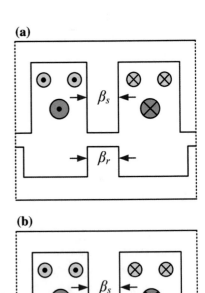

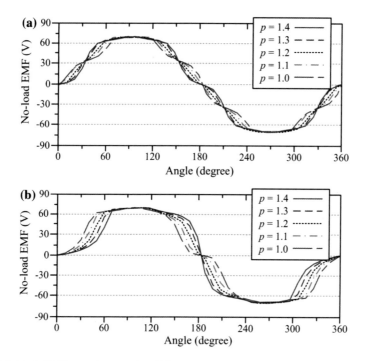

Fig. 10.8 Back EMF waveforms under different pole-arc ratios. **a** MTBF operation. **b** STUF operation

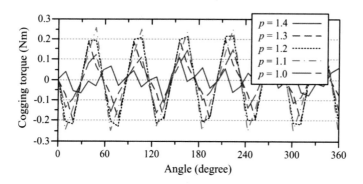

Fig. 10.9 Cogging torques under different pole-arc ratios

To confirm the pole-arc ratio with the optimal performance, the cogging torque is also analyzed and its waveforms under different value of p are shown in Fig. 10.9. In the case when $p = 1.2$ and 1.3, the peak value of the cogging torques are approximately 0.23 and 0.20 Nm, respectively. Undoubtedly, the lower cogging torque value always implies the better machine performance. Thus, it can be confirmed that the pole-arc ratio is optimized at $p_{\_opt} = 1.3$.

10.3.3   No-Load EMF Performances

With the support of the FEM, the no-load EMF waveforms of the proposed EG machine at the MTBF operation under the operating speed of 300 rpm and at STUF operation of 600 rpm are shown in Figs. 10.10 and 10.11, respectively. It can be shown that the no-load EMF waveforms at MTBF operation are well balanced with the three-phase symmetrical pattern, and hence confirming the effectiveness of the proposed balance-position winding arrangement. Meanwhile, the no-load EMF waveforms at STUF operation are also well balanced with the six-phase symmetrical pattern. The no-load EMF waveforms at both modes exhibit the patterns in between the sinusoidal-like and trapezoidal-like characteristics. Hence, both of them are suitable for both BLAC and BLDC conduction schemes.

Even though the operating speeds of two modes are different, the magnitudes of the no-load EMF between both modes are approximately the same. Hence, the results verify the MTBF operation should be adopted at the low-speed environment, whereas the STUF operation instead at the high-speed environment.

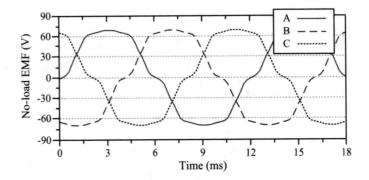

Fig. 10.10 Back EMF waveforms under MTBF operation at 300 rpm

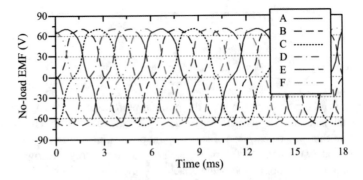

Fig. 10.11 Back EMF waveforms under STUF operation at 600 rpm

10.3.4 Torque Performances

The torque performances of the proposed machine at the MTBF operation and the STUF operation are carefully studied. As discussed, the BLAC and the BLDC conduction schemes, i.e., the three-phase sinusoidal-like and the six-phase trapezoidal-like currents, are suggested to employ in the MTBF and the STUF operations, respectively. This situation is so-called as the *Type I* and the torque performances are shown in Fig. 10.12. To provide the more comprehensive studies, the reversed situation or so-called as the *Type II*, i.e., the MTBF operation is employed with the three-phase BLDC scheme while the STUF operation with the six-phase BLAC scheme, is also included and as shown in Fig. 10.13.

It can be shown that the average steady torques at the MTBF and the STUF operations in *Type I* are around 9.9 and 5.6 Nm, respectively; while in *Type II* are around 10.8 and 5.0 Nm, respectively. The results show that the MTBF operation

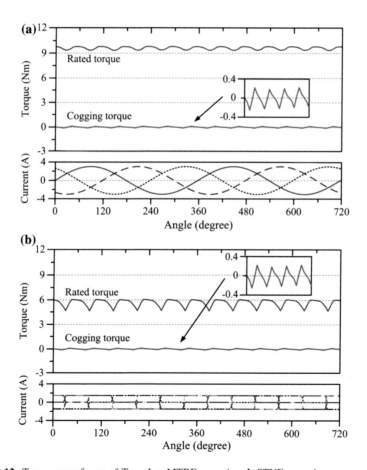

Fig. 10.12 Torque waveforms of Type I. **a** MTBF operation. **b** STUF operation

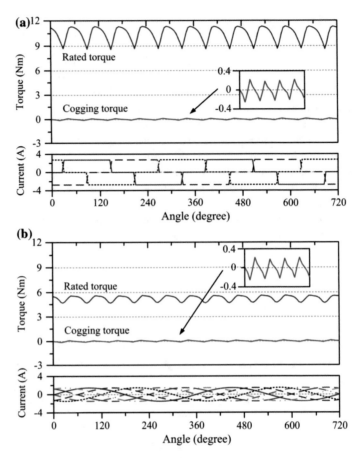

Fig. 10.13 Torque waveforms of Type II. **a** MTBF operation. **b** STUF operation

can achieve higher torques than those produced by the STUF operation. Meanwhile, it is also shown for the steady torque at the STUF operation, which produced based on the BLDC scheme, is larger than those on the BLAC does.

In addition, the peak value of the cogging torque is found to be approximately 0.2 Nm, which are only 2.1 and 3.8% of their corresponding average torques at the MTBF and the STUF operations, respectively, in *Type I*. Meanwhile, in *Type II*, the cogging torques are 1.9 and 4.2% of their corresponding average torques instead. All the cogging torque values are perceived as very acceptable, as compared with the profound PM counterparts [12].

To provide the more comprehensive analysis of the torque performances, the torque ripple values at the MTBF and the STUF operations in *Type I* are found to be 9.1 and 25.9%, respectively; while in *Type II* are 28.6 and 12.2%, respectively. The results have confirmed the torque ripple at the MTBF operation, which produced based on the BLAC mode, is smaller than those based on the BLDC does.

Table 10.2 Torque performances of the proposed EG machine

Item	Type I		Type II	
	MTBF	STUF	MTBF	STUF
Conduction scheme	BLAC	BLDC	BLDC	BLAC
No. of phases	3	6	3	6
Average torque (N m)	9.9	5.6	10.8	5.0
Cogging torque	2.1%	3.8%	1.9%	4.2%
Torque ripple	9.1%	25.9%	28.6%	12.2%

Similarly, two more combinations between the operation modes and the conduction schemes can be achieved, while they are less important. Hence, only the *Type I* and *Type II* are discussed and tabulated in Table 10.2. According to obtained results, the proposed machine should apply the *Type I* operation, i.e., the MTBF operation should adopt the BLAC scheme to produce relatively smoother torque for the high-torque low-speed operation, while the STUF operation instead adopt the BLDC scheme to achieve relatively better torque density for the low-torque high-speed operation.

10.3.5 Operating Range Extension Performances

The torque-speed characteristic of the proposed machine is shown in Fig. 10.14. In order to maintain the high levels of the DC-field excitation, different operations are suggested for certain operating ranges as

1. In 0–300 rpm, the MTBF operation should be employed with the rated DC-field excitation, i.e., 5 A/mm^2.

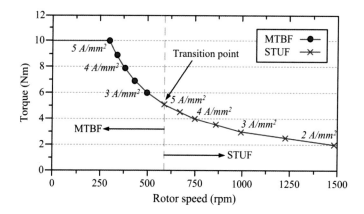

Fig. 10.14 Torque-speed characteristics of EG machine

2. In 300–600 rpm, the MTBF operation should be employed based on the flux-weakening operation.
3. At 600 rpm, the proposed machine should switch from the MTBF operation to the STUF operation, while the DC-field excitation is regulated to its rated value, i.e., 5 A/mm^2.
4. In 600–1200 rpm, the STUF operation should be employed based on the flux-weakening operation.

With the incorporation of the EG and flux-weakening operations, the proposed machine can maintain at the high levels of DC-field excitation. This can prevent the machine to switch to operate as a switched reluctance machine, and hence minimizing the unwanted consequences. Moreover, the proposed machine can cover the whole range from 0 to 1200 rpm, and therefore fulfilling the requirements of the direct-drive applications for EVs.

10.4   Experimental Verifications

To verify the proposed idea and the predicted machine performances, the experimental prototype of the proposed dual-mode EG machine is developed and as shown in Fig. 10.15. The corresponding key design data of the proposed EG machine is listed in Table 10.3.

Fig. 10.15 Experimental setup of EG machine. **a** Stator. **b** Rotor

Table 10.3 Key design data of the proposed EG machine

Items	Value
Stator outside diameter (mm)	156.0
Stator inside diameter (mm)	96.0
Rotor outside diameter (mm)	95.0
Rotor inside diameter (mm)	22.0
No. of stator poles	12
No. of rotor poles	10
Stator pole arc (°)	12.0
Rotor pole arc (°)	15.6
Airgap length (mm)	0.5
Stack length (mm)	120
No. of turns per armature coil	110

Fig. 10.16 Measured no-load EMF waveforms at MTBF mode

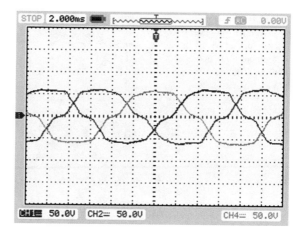

The measured no-load EMF waveforms of the proposed dual-mode EG machine at the MTBF operation under the operating speed of 300 rpm and at STUF operation of 600 rpm are shown in Figs. 10.16 and 10.17, respectively. Because all the no-load EMF waveforms among the six-phase windings at STUF mode are well balanced with symmetrical patterns, to achieve the better presentation, only the A-phase, C-phase and E-phase are shown. As confirmed, the measured waveforms well align with the simulated results as shown in Figs. 10.10 and 10.11, respectively, while the slightly differences shown in the results are generally caused by the end-effect and manufacturing imperfection. In the meantime, the measured magnitudes of the no-load EMF waveforms well comply with the theoretical ones, so that the minor discrepancies are believed to be acceptable.

In addition, the measured no-load EMF waveforms at MTBF mode and at STUF mode with higher operating speeds, without and with the DC-field flux regulations, are shown in Figs. 10.18 and 10.19, respectively. With the flux-weakening

Fig. 10.17 Measured
no-load EMF waveforms of
A-phase, C-phase, and
E-phase at STUF mode

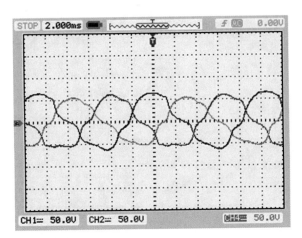

Fig. 10.18 Measured
no-load EMF waveforms of
EG machine at MTBF mode
with higher operating speed.
a Without flux regulation.
b With flux regulation

(a)

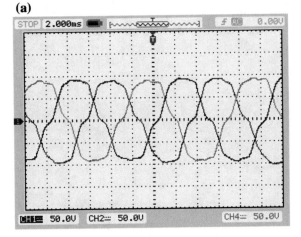

(b)

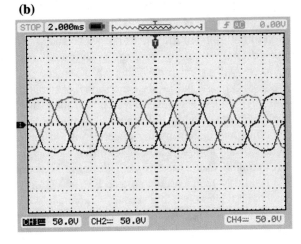

Fig. 10.19 Measured
no-load EMF waveforms of
EG machine at STUF mode
with higher operating speed.
a Without flux regulation.
b With flux regulation

(a)

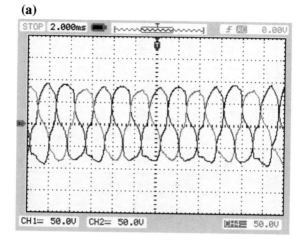

(b)

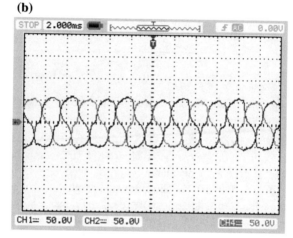

capability, the magnitudes of the no-load EMF waveforms at higher speed ranges
can be maintained at the same level as those with the lower speed ranges. These
measured waveforms confirm the expected flux regulating capability offered by the
proposed EG machine. Consequently, these measured values can verify the proposed system can be operated over a wide range of operating speeds.

Upon the rectifications, the corresponding simulated and measured voltage
characteristics at two modes under DC-field excitation of 3 A, with respect to the
operating speed at no-load conditions are shown in Fig. 10.20. As illustrated, the
simulated and measured results are in good agreement, where the generated voltages increase linearly with the operating speed. Furthermore, the simulated and

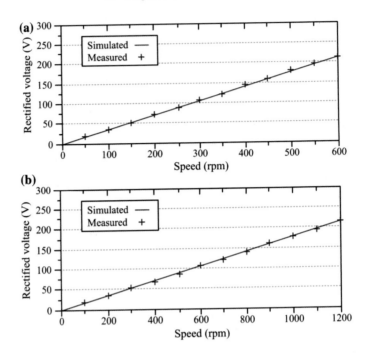

Fig. 10.20 Simulated and measured output voltage characteristics at different operating speeds. **a** MTBF mode. **b** STUF mode

measured voltage characteristics at MTBF mode under operating speed of 300 rpm and at STUF mode under 600 rpm, with respect to the DC-field excitation are shown in Fig. 10.21. The simulated results align with the measured results, where the generated voltages can be regulated linearly based on the controllable DC-field excitations without severe saturation. The results verify that the proposed EG machine can utilize the mode-changing and flux regulating capabilities to maintain the output voltages at certain value, and hence protecting the whole EV system.

The corresponding simulated and measured torque characteristics at two modes under DC-field excitation of 2 A, with respect of the armature current are shown in Fig. 10.22. Due to the end-effect, the measured torques are slightly smaller than the simulated one. Yet, good overall agreement is achieved. According to the measured results regarding the back EMF and torque performances, the MTBF mode should be employed as the high-torque low-speed operation, while the STUF mode as the low-torque high-speed operation. Hence, the concept of the proposed EG machine is verified.

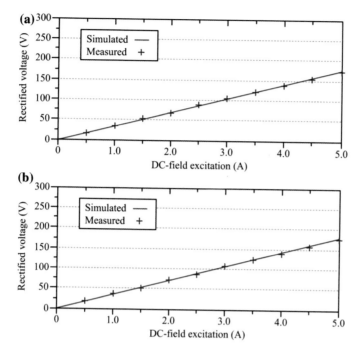

Fig. 10.21 Simulated and measured voltage characteristics at different DC-field currents. **a** MTBF mode under 300 rpm. **b** STUF mode under 600 rpm

Finally, the efficiencies of the proposed EG machine at two modes as the function of load currents have been measured and shown in Fig. 10.23. In particular, the efficiencies of the proposed machine at MTBF mode and at STUF mode can achieve approximately 78 and 74%, respectively. According to the experimental results, the proposed machine at both modes can achieve satisfactory efficiencies as compared with the commonly employed machines do. To be specific, the efficiency of the magnetless switched reluctance (SR) machine, the DSPM machine and the FSPM machine are around 70% [13], 80% [14], and 85% [7], respectively. Upon the installation of the high-energy-density PM materials, the DSPM and the FSPM types can provide higher efficiencies as well as higher torque densities than the magnetless counterparts do. However, the magnetless counterparts instead enjoy the definite advantage of better cost-effectiveness than the PM candidates. In the meantime, with the incorporation between the mode-changing and the flux regulating capability, the proposed EG machine enjoys the highest flux controllability as well as widest operating range among the commonly employed candidates do. For better illustration, the comparisons on these machines are summarized in Table 10.4.

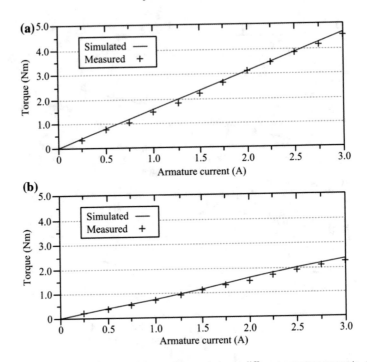

Fig. 10.22 Simulated and measured torque characteristics at different armature currents. **a** MTBF mode. **b** STUF mode

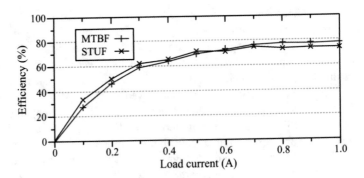

Fig. 10.23 Measured efficiencies of EG machine at different load currents

Table 10.4 Comparisons among commonly employed machines

Items	SR	DSPM	FSPM	EG
Efficiency	Low	Moderate	High	Moderate
Power density	Low	Moderate	High	Moderate
Cost-effectiveness	Moderate	Low	Low	High
Flux controllability	Low	Low	Low	Very high
Operating range	Narrow	Narrow	Narrow	Wide

10.5 Summary

This chapter introduces a new dual-mode EG machine for the EV applications. The balance-position winding configuration is proposed to apply in the dual-mode machine to improve its flux-linkage patterns. In addition, the selection criteria of the no-load EMF waveforms are also carefully discussed in order to improve the machine performances at the extreme situations. Namely, the proposed machine can offer the smoother torque at the low-gear situation and the better torque density at the high-gear situation. By the mode-changing principle and the DC-field regulation, the proposed machine can effectively extend its operating range, to fulfill the requirements for the EV applications.

References

1. C.C. Chan, The state of the art of electric, hybrid and fuel cell vehicles. Proc. IEEE **95**(4), 704–718 (2007)
2. A.Y. Saber, G.K. Venayagamoorthy, Plug-in vehicles and renewable energy sources for cost and emission reductions. IEEE Trans. Industr. Electron. **58**(4), 1229–1238 (2011)
3. Z.Q. Zhu, D. Howe, Electrical machines and drives for electric, hybrid and fuel cell vehicles. Proc. IEEE **95**(4), 746–765 (2007)
4. A. Tenconi, S. Vaschetto, A. Vigliani, Electrical machines for high-speed applications: design considerations and tradeoffs. IEEE Trans. Industr. Electron. **61**(6), 3022–3029 (2014)
5. C. Yu, K.T. Chau, Design, analysis, and control of DC-field memory motors. IEEE Trans. Energy Convers. **26**(2), 479–489 (2011)
6. J.T. Chen, Z.Q. Zhu, Winding configurations and optimal stator and rotor pole combination of flux-switching PM brushless AC machines. IEEE Trans. Energy Convers. **25**(2), 293–402 (2010)
7. R.P. Deodhar, A. Pride, S. Iwasaki, J.J. Bremner, Performance improvement in flux-switching PM machines using flux diverters. IEEE Trans. Ind. Appl. **50**(2), 973–978 (2014)
8. C.H.T. Lee, K.T. Chau, C. Liu, C.C. Chan, Overview of magnetless brushless machines. IET Electr. Power Appl. (to appear)
9. I. Boldea, L.N. Tutelea, L. Parsa, D. Dorrell, Automotive electric propulsion systems with reduced or no permanent magnets: an overview. IEEE Trans. Industr. Electron. **61**(10), 5696–5711 (2014)
10. C.H.T. Lee, K.T. Chau, C. Liu, Design and analysis of a dual-mode flux-switching doubly salient DC-field magnetless machine for wind power harvesting. IET Renew. Power Gener. **9**(8), 908–915 (2015)

11. Z.Q. Zhu, J.T. Chen, Y. Peng, D. Howe, S. Iwasaki, R. Deodhar, Analysis of a novel multi-tooth flux-switching PM brushless AC machine for high torque direct-drive applications. IEEE Trans. Magn. **44**(11), 4313–4316 (2008)
12. M. Cheng, W. Hua, J. Zhang, W. Zhao, Overview of stator-permanent magnet brushless machines. IEEE Trans. Industr. Electron. **58**(11), 5087–5101 (2011)
13. C. Lee, R. Krishnan, N.S. Lobo, Novel two-phase switched reluctance machine using common-pole E-core structure: concept, analysis, and experimental verification. IEEE Trans. Ind. Appl. **45**(2), 703–711 (2009)
14. Y. Fan, K.T. Chau, M. Cheng, A new three-phase doubly salient permanent magnet machine for wind power generation. IEEE Trans. Ind. Appl. **42**(1), 53–60 (2006)

Chapter 11
Conclusions and Future Works

11.1 Conclusions

The objectives of this project have been successfully accomplished and various high-performance magnetless doubly salient brushless machines have been analyzed and developed. The proposed magnetless machines enjoy the definite advantages of high cost-effectiveness, wide operating range, flexible flux controllability and improved torque performance. In particular, some of the proposed concepts have been published as the referred journal papers and conference papers.

As compared with the permanent-magnet (PM) brushless machines, the magnetless candidates undoubtedly suffer from relatively lower torque performances. To relieve the situation, the multi-tooth structure described in Chap. 3 is thoroughly analyzed to form the corresponding machines with improved performances. With the flux-modulation effect, the multi-tooth machines can result higher torque densities. In the meantime, the concept of multi-tooth structure has been further extended to form the double-rotor (DR) machines described in Chap. 4 for some special applications.

Apart from the torque density, the torque ripple value is another most important criteria to determine the electric machine performances. The concept of the mechanical-offset (MO), which can integrate the pulsating torque components seamlessly, has been proposed in Chap. 3. However, the original design suffers from the problems of control complexity and higher power electronics costs. To improve the situation, the singly fed mechanical-offset (SF-MO) machine has been proposed in Chap. 5.

To increase the market penetration as well as product attractiveness of the magnetless brushless machines, the proposed concepts have been quantitatively compared with the commonly employed candidates. To be specific, the developed flux-reversal DC-field (FRDC) machine described in Chap. 6 and the axial-field (AF) machine described in Chap. 8 can both offer outstanding cost-effectiveness with satisfactory torque performances, as compared with the induction machine and

© Springer Nature Singapore Pte Ltd. 2018
C. H. T. Lee, *Design, Analysis and Application of Magnetless Doubly Salient Machines*, Springer Theses,
https://doi.org/10.1007/978-981-10-7077-8_11

the PM counterparts. As confirmed, the proposed magnetless machines have shown great potential in various applications.

With the utilization of the identical torque properties between the two side rotors in the AF machines, based on the foundation in Chap. 8, the magnetic steering (MS) machines are developed to realize the magnetic differential (MagD) system for EV application described in Chap. 9. Based on the highly reliable magnetic interlocking mechanism, the proposed MagD system has shown distinct merits over the conventional electronic differential (ED) system.

By purposely combining the design equations of two machines, the new types of machine, namely the dual-mode machine described in Chap. 7 and the electronic-geared (EG) machine described in Chap. 10 have been proposed. With the reconfiguration of winding arrangements, the proposed machines can switch between two modes to cater different situations. Hence, as compared with the existing candidates, the proposed machines can offer wider operating ranges with higher control flexibility.

Despite there are numerous magnetless brushless machines are proposed, four of them with the most attractive performances, namely the SF-MO machine (Chap. 5), the FRDC machine (Chap. 6), the dual-mode machine (Chap. 7) and the EG machine (Chap. 10) have been developed as the experimental prototypes. All the experimental setups are developed from the conceptual stage, while the corresponding processes, including the machine simulation, performance analysis, request of tender from manufacturers, installation of testbed and experimental testing have been thoroughly conducted. The gained experiences have been recorded as the guidelines for the upcoming research members as reference.

The proposed magnetless brushless machines have been purposely designed based on the typical requirements of various industrial applications, such as the electric vehicle (EV) application and the wind power generation. In particular, with the high cost-effectiveness and excellent stability, the FRDC machine, the dual-mode machine and the EG machine have shown great potential for the mentioned applications.

11.2   Future Works

Despite all the objectives of the project have been successfully achieved, there are still areas for further developments as follows

1. In this project, the machine performances are mainly analyzed based on the steady stage analysis. Even though most of the important performances can be realized by the steady stage analysis, the transient stage performances should be developed to provide the more comprehensive analysis.
2. In the practical applications, the thermal characteristic is one of the most important issues that should be addressed. In this project, the dimensions of the magnetless machines are optimized so that the core losses are minimized

accordingly. Hence, the potential problems caused by the thermal issue are relieved. However, the static and transient thermal analyses of the magnetless machines are worth developing.

3. The mechanical vibration and acoustic noise produced by the proposed machines can definitely influence the system stability, and hence these problems should be carefully studied. In general, the vibrations as well as noises are generated by the electromagnetic force during the machine operation, while the researches on the minimization or utilization of the undesirable vibrations are interesting.

4. The fundamental machine characteristics and performances have been analyzed and given in this project, while the development of the advanced control algorithms is a very interesting research topic for the magnetless brushless machines. Unlike the traditional candidates, the newly developed magnetless machines consist of an additional DC-field excitation, so that the advanced control schemes with higher flexibility or higher efficiency can be potentially achieved.

Printed in the United States
By Bookmasters